NASA: Politics above Science

J. Marvin Herndon, Ph.D.

NASA: Politics above Science

Copyright © 2018 by J. Marvin Herndon, Ph.D.

All rights reserved. No part of this book may be reproduced or transmitted in any form or by any means without written permission of the author.

Back cover image courtesy of David Tulis.

Table of Contents

Chapter 1. Politics First and Foremost 1

Chapter 2. Misrepresentation and Deceit 11

Chapter 3. Nuclear Reactor Debacle 27

Chapter 4. NASA's Faulty Storyline 73

Chapter 5. Science NASA Missed 87

Chapter 6. Science NASA Bungled 95

Chapter 7. Deception at NASA Headquarters 105

Appendix: Corruption of Science in America 111

References 135

Chapter 1. Politics First and Foremost

"I believe that this Nation should commit itself to achieving the goal, before this decade is out, of landing a man on the Moon and returning him safely to the Earth. No single space project in this period will be more impressive to mankind or more important for the long range exploration of space. And none will be so difficult or expensive to accomplish."

Figure 1. U. S. President John F. Kennedy addressing a joint session of the U. S. Congress on May 25, 1961.

With those challenging words to a joint session of Congress on May 25, 1961, President John F. Kennedy (Figure 1) set the course for the fledgling National Aeronautics and Space Administration (NASA) that had been created by an Act of Congress not quite three years before.

On July 20, 1969, Apollo 11 astronauts Neil Armstrong and Edwin "Buzz" Aldren, Jr. landed the Lunar Module Eagle on the Moon. The next day, these two astronauts became the first humans to walk on the Moon (Figure 2). Eagle spent 21½ hours on the lunar surface while astronaut Michael Collins remained in lunar orbit with the Command/Service Module Columbia. On July 24, 1969 the three Apollo 11 astronauts returned safely to Earth with a splash-down in the Pacific Ocean.

Figure 2. July 21, 1969. Neil Armstrong, the first human to set foot on the Moon, beside the Lunar Module "Eagle" on the Moon's surface.

The safe return of the astronauts from their first mission to the Moon marked the successful achievement of the goal pronounced by President Kennedy eight years before, and was the highpoint of NASA's golden era. The lunar landing vaulted the United States to first place in the race to space that was begun in 1957 by the U.S.S.R. with the successful orbit of Earth's first artificial satellite, Sputnik 1. Landing the first humans on the Moon was a technological marvel, a historical milestone, and, especially, was a source of great pride as just about every American who was able watched the event live on television. At that time, could anyone have imagined that America's 'can do' space agency in just two decades would become an agency of spectacular and disastrous blunders and worse?

In the nearly two decades following NASA's best result of landing men on the Moon and returning them safely to Earth, a new organizational culture had taken root at NASA. The collective behavior and commonality of values of individuals associated with and unique to NASA, the new organizational culture, no longer was directed by scientific and engineering ethos, but by political mentality and public relations stunts. Spacecraft were to a large extent simply the vehicles to achieve NASA's new political agenda.

NASA's shift in organizational culture was clearly evident when, on April 12, 1985, at great expense to taxpayers, the Space Shuttle Discovery launched with U. S. Senator Jake Garn (Figure 3 left) onboard designated as a "Payload Specialist". This was clearly a political joy-ride; Senator Garn was a member of the

Senate Appropriations Committee and the Chairman of a sub-committee that dealt with NASA.

Figure 3. Political Astronauts. Left, U. S. Senator Jake Garn. Right, Schoolteacher Christa McAuliffe.

Even before Garn's joy-ride, beginning in 1984, plans were being made for the greatest public relations stunt yet of NASA's new culture, the Teacher in Space Project. After a much publicized search, high school social studies teacher Christa McAuliffe (Figure 3 right) was selected for the Challenger Space Shuttle flight that was originally scheduled to launch from Kennedy Space Center in Florida mid-afternoon of January 22, 1986. But unknown to McAuliffe and the crew, the Shuttle had a potentially fatal design flaw that NASA management had known about since 1977 but had failed to address. And, there were the delays.

Delays in the previous mission caused the Challenger launch to be rescheduled to January 23 and then to January 24. Bad weather at the Transoceanic Abort Landing site in Dakar, Senegal, caused the schedule to be further delayed until January 25, and the launch time to be moved to morning to accommodate an alternative Abort Landing site at Casablanca that was not equipped for night operations. But severe weather in Florida caused still further postponement until the morning of January 27. But then the launch was delayed yet another day by problems with the exterior access hatch, first a malfunctioning indicator, then a stripped bolt, and then more delays because of strong crosswinds.

The night before the January 28 morning launch of Challenger was a time of great concern at Morton Thiokol, the manufacturers of the two solid fuel rockets that help boost Challenger off the launch pad. Each of the solid fuel rockets was assembled from six pieces with three factory-sealed joints and three field-assembled joints, each sealed with two rubber O-rings. Tests made in 1977 showed that pressures could bend the metal parts exposing the rubber gaskets to hot exhaust which could compromise the joint and lead to disaster. Engineers at Morton Thiokol knew that the freezing weather would make the rubber O-rings brittle and even more prone to problems, but against their advice, NASA management pressed for launch.

Upon launch, the O-rings failed immediately, but the flame channels were temporarily blocked by the aluminum oxide rocket debris. But then, 72 seconds into flight, the temporary seal broke loose and the solid rocket flames breached the O-ring and detonated the liquid hydrogen and oxygen tanks (Figure 4). All aboard Challenger perished.

Figure 4. Destruction of the Space Shuttle Challenger about 76 seconds after launch.

In the aftermath, the U. S. House Committee on Science and Technology, which conducted hearings and reviewed the report of the Presidential Commission on the Space Shuttle Challenger Accident, reported in part "...the Committee feels that the underlying problem which led to the Challenger accident was not poor communication or underlying procedures as implied by the Rogers Commission conclusion. Rather, the fundamental problem was poor technical decision-making over a period of several years by top NASA and contractor personnel, who failed to act decisively to solve the increasingly serious anomalies in the Solid Rocket Booster joints."

On February 1, 2003, another disaster: The Space Shuttle Columbia disintegrated upon re-entry into the Earth's atmosphere (Figure 5). During previous shuttle launches, pieces of foam insulation had frequently been observed shedding from the exterior of the main fuel tank and had upon occasion scarred the protective heat shield, but NASA management considered potential damage to be an acceptable risk, just as the O-rings had been an acceptable risk seventeen years ago before the Challenger disaster. This time, though, the foam insulation debris struck and fatally damaged the leading edge of Columbia's left wing heat shield during launch. All astronauts on board perished as Columbia disintegrated during re-entry.

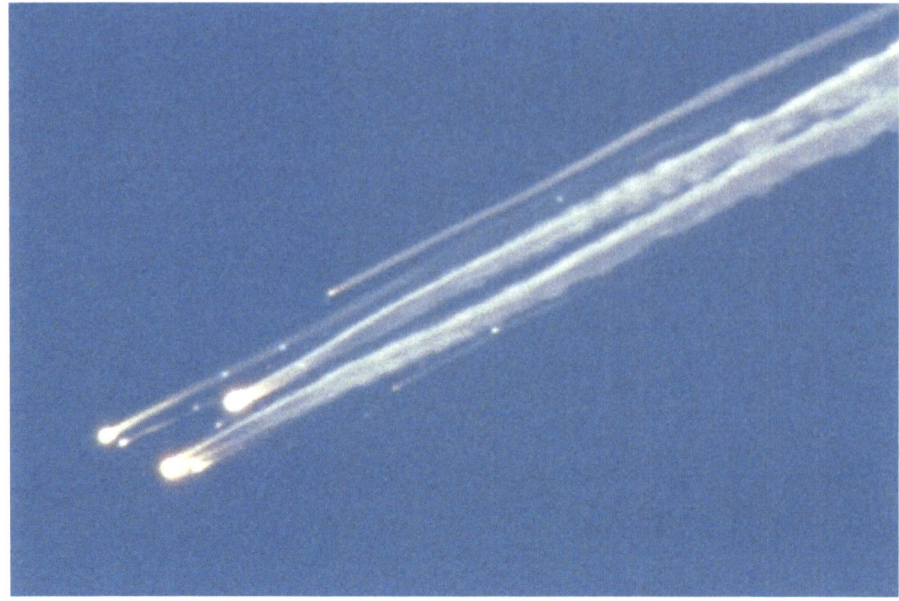

Figure 5. The Space Shuttle Columbia disintegrated over Texas during re-entry.

The Columbia Accident Investigation Board Report stated: Management decisions made during Columbia's final flight reflect missed opportunities, blocked or ineffective communications channels, flawed analysis, and ineffective leadership. Perhaps most striking is the fact that management … displayed no interest in understanding a problem and its implications…. In fact, their management techniques unknowingly imposed barriers that kept at bay both engineering concerns and dissenting views, and ultimately helped create "blind spots" that prevented them from seeing the danger the foam strike posed.

Reports of investigations of the two Space Shuttle disasters disclose the poor technical decision-making indicative of NASA's politically-oriented, organizational culture. These two tragic events, however, are only the highly visible markers. Out of sight and far more persistent and pervasive is NASA's concerted misrepresentation and deception of planetary science, NASA's systematic failure to tell the full truth.

Telling the full truth is what science is all about. Indeed, truth is the pillar of civilization. The word 'truth' occurs 224 times in the King James Version of the Holy Bible; witnesses testifying in American courts and before the United States Congress must swear to tell the truth; and, laws and civil codes require truth in advertising and in business practices. At the United States Naval Academy, truth is an important component of that institution's Honor Concept: "Midshipmen are persons of integrity: They stand for that which is right. They tell the truth and ensure that the full truth is known. They do not lie...." Scientists engaged in investigations for NASA, however, adhere to a different standard, an extension and perversion of university-spawned 'politically correct' speech which ensures that the full truth is not revealed.

NASA policies and lack of oversight by NASA officials has led, I allege, to institutional corruption within the NASA establishment that has persisted for at least two decades, and that has diminished NASA's scientific credibility and tarnished America's integrity. As disclosed here, for decades NASA officials have permitted secret reviews by competitors, thus limiting competition for grants and contracts, and intimidating NASA scientists into never mentioning non-NASA new, relevant concepts, published in world-class scientific journals, that contradict the NASA storyline, thus failing to reveal the full truth in scientific investigations undertaken at taxpayer expense, deceiving the scientific community, teachers, students, and the general public and trivializing scientific investigations.

As further revealed, for decades NASA officials have allowed the widespread making of models, which are based upon arbitrary assumptions and which fail to cite previous and/or contradictory work, instead of making scientific discoveries. There is a particularly dark side to making models: Models based upon arbitrary assumptions are (wrongly) promoted as being scientific and the absence of scientific standards attending their use can lead to gross misrepresentation such as was experienced with "climate models" in the global warming scare that began to unravel with the Climategate revelations.

This book describes from a scientific standpoint NASA's involvement in the global warming scare, NASA's deception regarding planetary origins and the internal energy and magnetic field generation in planets and large moons, NASA's bungling of Project Stardust, NASA's bungling of the Mercury MESSENGER results which made it impossible for MESSENGER scientists to explain one of its most important observations, and NASA's failure to investigate serious allegations of wrong doing. This is the true story of that NASA deception, from bottom to top, and the scientific advances that NASA failed to make as a consequence.

Chapter 2. Misrepresentation and Deceit

Prior to World War II, a graduate student working on a Ph.D. degree was expected to make a new discovery to earn that degree, even if it meant starting over after years of work because someone else made the discovery first. During that period, scientific integrity and self-discipline were the hallmarks of a scientist. I earned my Ph.D. in nuclear chemistry in 1974 and was immediately invited to begin a three post-doctoral apprenticeship with Nobel Laureate Harold C. Urey, who had served a post-doctoral apprenticeship with Nobel Laureate Niels Bohr, and with Hans E. Suess, grandson of famed geologist Eduard Suess and co-discoverer of the shell structure of the atomic nucleus for which his colleague Hans Jensen was awarded the Nobel Prize.

Figure 6. Nobel Laureate Harold C. Urey (1893-1981) on left and Hans E. Suess (1909-1993).

Urey and Suess (Figure 6) were rigorously trained in science prior to World War II and chose to pass along to me some of what they had learned, including knowledge of how to make scientific discoveries. Their timing was particularly opportune, as the influx of grant money from NASA and the National Science Foundation (NSF), coupled with fundamental science management flaws, such as use of secret reviews by one's competitors [1] [see Appendix], had already begun to infect NASA and the scientific community with the questionable practice of making models based upon assumptions, instead of making scientific discoveries. Why? Virtually anyone can make a model of some imagined event or process, beginning with assumptions and tying together bits and pieces with impressive-looking equations that might or might not be relevant. Models do not have to be correct and model makers, I have observed, generally have no inclination to adhere to long-standing scientific standards, such as citing relevant prior discoveries or published contradictions. About models, the American statistician, George E. P. Box, has stated: "Essentially, all models are wrong, but some are useful" [2].

Teachers of laboratory courses in high school and college rarely tell students beforehand the value they are trying to determine by experiment, because too often the student will manipulate the data to arrive at that known result. In making models, scientists know beforehand what they are modeling and what they think the result is supposed to be or what they want the result to be. So, the temptation is for model-makers to adjust their assumptions and model parameters, and result-

select or agenda-select data to achieve the known-beforehand result. NASA-funded scientists like to make models based upon arbitrary assumptions and make models based upon other models. Even if the model appears to model what is intended to be modeled, there is no certainty that nature behaves the way being modeled. Models do not have to be correct and usually are not. Making models, which many can do easily, is no substitute for the much more difficult task of making scientific advances by discovering fundamental quantitative relationships in nature that are securely anchored to the properties of matter. Models often become part of the NASA "accepted" storyline, which can lead to widespread confusion about the science NASA purports to investigate.

Notice that as you heat a pot of water on the stovetop, before it starts to boil, the water begins to circulate from bottom to top and from top to bottom. This is called convection and it can be better observed by adding a few tea leaves, celery seeds, or the like, which are carried along by the circulation of water. Convection occurs because heat at the bottom causes the water to expand a bit, becoming lighter, less dense, than the cooler water at the top. This process of convection is an unstable, top-heavy arrangement which attempts to regain stability by fluid motions.

In 1931, the British geologist Arthur Holmes (1890-1965) suggested that Earth's mantle, the 2900 km thick, solid-rock shell beneath the crust was convecting [3]. He thought that mantle convection might explain Alfred Wegener's idea that the continents are drifting apart. In the 1960's when plate tectonics was being formulated, mantle convection was adopted as a crucial component. Then along came the model-makers spending untold millions of dollars on mantle convection models, computer programs that purported to describe mantle convection. All of the models began with the assumption that mantle convection exists and then assumed parameters and methods that would supposedly lead to mantle convection. In other words, if one set out to model mantle convection, that is what the model must show. Some used invalid mathematics; the most sophisticated models, though, assumed that the solid-rock mantle behaves an as ideal gas acting without frictional loss, the rock assumed to ever-expand as it moves outward against no resistance. But the earth's mantle is not an ideal gas. One reason we know this is that at depths up to 660 km earthquakes occur, which represent the catastrophic release of pent-up stress; an ideal gas has no stress [4].

Mantle convection models were made and re-made, and books were written about the models without any one asking the question: "What's wrong with this picture?" But I did, and realized that because of the weight above, the rock at the bottom of the Earth's mantle is about 62% more dense than at the top. The small amount density-decrease caused by thermal expansion at the bottom, less than 1%, is insufficient to make the bottom-heavy mantle top-heavy. Mantle convection is physically impossible [4]. Being trapped in a model morass, NASA scientists extended mantle convection models to Mercury and Mars.

There is a very serious and particularly dark side to making models: Models based upon arbitrary assumptions are (wrongly) promoted as being scientific and generally do not adhere to long-existing standards of ethical science, such as citing contradictory evidence. The absence of scientific standards attending their use can lead to gross misrepresentation such as was experienced with "climate models" in the global warming scam that began to unravel with the Climategate revelations.

In 1957, Roger Revelle and Hans E. Suess published a scientific paper, entitled "Carbon dioxide exchange between atmosphere and ocean and the question of an increase of atmospheric CO2 during the past decades." [5]. In that paper Revelle and Suess suggested that the Earth's oceans would absorb excess carbon dioxide generated by humanity at a much slower rate than previously predicted by geoscientists and might create a "greenhouse" effect that might eventually lead to global warming.

Twenty years after publication of the Revelle-Suess paper, I had lengthy discussions with Suess about the possibility of global warming and the necessity to validate the effect of increased atmospheric carbon dioxide in a broader framework that includes the potential variability Earth's major heat source, the Sun, the factors potentially affecting the variability of Earth's reflectivity and its consequences, the potential variability of heat being brought to the surface from deep within the Earth and its potential impact on ocean warming and on changes in ocean circulation. A vast amount of high-quality data has been obtained since the time of those discussions, for example, oxygen isotope paleo-temperatures from Antarctic ice cores. But even at present, great unknowns, especially about potential variability of Earth's two primary energy sources, render impossible the making valid, unambiguous, scientific determinations on the possibility of anthropogenic (human caused) global warming.

Then along came the climate modelers, with grand "climate models", based upon the false assumptions that heat from the Sun and heat from within the Earth are both constant. With those predominant variables unrealistically held constant, the tiny greenhouse effect of carbon dioxide appears significant [6]. The intended result of those climate models is to demonstrate that human activities are indeed causing global warming and that the consequences are planet-threateningly dire. Press coverage abounded: disaster scenarios sell magazines and newspapers. Fear of global warming created political and financial opportunities. The public coffers opened and the gravy train left the station. Climate modelers were riding high with burgeoning grant resources, greater visibility, and conferences in exotic places, like Bali. At the top of the heap, former U. S. Vice President Al Gore was scaring school children with An Inconvenient Truth, while reaping vast income from the movie, lectures, and commercial ventures. Similarly, the Climate Research Unit at the University of East Anglia in the United Kingdom (CRU) was at the forefront of global warming climate models. Then, the great global warming scare began to unravel.

In November, 2009, over a thousand emails were leaked from an internal computer system within CRU; the incident became known as Climategate [7,8]. As noted in a November 24, 2009 Wall Street Journal article: "Yet even a partial review of the emails is highly illuminating. In them, scientists appear to urge each other to present a 'unified' view on the theory of man-made climate change while discussing the importance of the 'common cause'; to advise each other on how to smooth over data so as not to compromise the favored hypothesis; to discuss ways to keep opposing views out of leading journals; and to give tips on how to 'hide the decline' of temperature in certain inconvenient data.… When deleting, doctoring or withholding information didn't work, Mr. Jones suggested an alternative in an August 2008 email to Gavin Schmidt of NASA's Goddard Institute for Space Studies [GISS], copied to Mr. Mann. "The FOI [Freedom of Information] line we're all using is this," he wrote. 'IPCC is exempt from any countries FOI—the skeptics have been told this. Even though we . . . possibly hold relevant info the IPCC is not part of our remit (mission statement, aims etc) therefore we don't have an obligation to pass it on'....we do now have hundreds of emails that give every appearance of testifying to concerted and coordinated efforts by leading climatologists to fit the data to their conclusions while attempting to silence and discredit their critics."

After Climategate, almost everywhere the phrase "global warming" was replaced with "climate change", but the countless agenda-driven scientists, as well as the political, scientific, and educational organizations that promoted global warming, are not apt to relinquish quickly the non-science associated with the global warming agenda.

So, what was NASA's role in the global warming deception? And what should NASA's role have been? Science is about telling the truth, the full truth. With its global temperature sensing satellites, NASA should have maintained the highest standard of scientific integrity and objectivity. Instead, NASA did the opposite. NASA became a willing contributor to the global warming deception. There is no surprise here; as such behavior is entirely in keeping with NASA's politically-based organizational culture that does not follow scientific standards.

Since 1979, NASA has been using satellites to measure air temperature of the lower troposphere directly above the Earth's surface. The global satellite temperature data has been validated by radiosonde weather balloon measurements. Instead of just relying upon thirty years of validated global coverage satellite data, the gatekeeper of NASA's climate data chose instead to use the problematic networks of land and ocean based sensors.

The NASA Goddard Institute for Space Studies (GISS), a component laboratory of NASA's Goddard Space Flight Center, conducts research that "emphasizes a broad study of Global Change, the natural and anthropogenic changes in our environment that affect the habitability of our planet" as its current home page states. From 1981 until 2013, GISS was directed by James E. Hansen, who essentially served as the de facto gatekeeper for NASA's climate related models and underlying data.

In science laboratory courses, students generally are taught the importance of objectivity and the necessity to eliminate bias. So, one might reasonably expect a NASA laboratory to be directed by an unbiased individual. But, that was not the case for NASA's GISS. James Hansen has a long and very public record of being a pro-global warming activist.

From an article in the Wall Street Journal Aug. 29, 2007: "What's more disturbing is what this incident tells us about the scientific double standard in the global warming debate. If this kind of error were made by climatologists who dare to challenge climate-change orthodoxy, the media and environmentalists would accuse them of manipulating data to distort scientific truth. NASA's blunder only became a news story after Internet bloggers played whistleblower by circulating the new data across the Web.

So far this year NASA has issued at least five press releases that could be described as alarming on the pace of climate change. But the correction of its overestimate of global warming was merely posted on the agency's Web site. James Hansen, NASA's ubiquitous climate scientist and a man who has charged that the Bush Administration is censoring him on global warming, has been unapologetic about NASA's screw up. He claims that global warming skeptics -- "court jesters," he calls them -- are exploiting this incident to "confuse the public about the status of knowledge of global climate change, thus delaying effective action to mitigate climate change."

So let's get this straight: Mr. Hansen's agency [NASA] makes a mistake in a way that exaggerates the extent of warming, and this is all part of a conspiracy by 'skeptics'?"

As noted in a January 1, 2009 *Heartlander Magazine* article by James M. Taylor: "Many climate scientists have criticized GISS in recent years for routinely claiming significantly higher global temperatures than those reported by other scientists; for employing a staff that appears to see its role more as advocates than as scientists; for getting caught claiming recent years were warmer than the data indicated; and for failing to provide transparency in how they manipulate raw temperature data before presenting their adjusted "official" temperature reports.

After GISS generated substantial media attention with its claim October 2008 was the warmest October in history, a number of global warming "skeptics" smelled something fishy and examined the data themselves. They soon discovered NASA and its partners at the National Oceanic and Atmospheric Administration had copied the September 2008 temperature data from Russia into the October Russian temperature dataset."

These examples reveal part of a broader pattern of NASA's questionable "global warming" reporting that "skeptics" have revealed. So, one might ask, why did NASA management not re-assign Hansen to a position where his bias would not appear to influence or distort NASA's science? Why? Perhaps because Hansen's activities helped to insure that NASA had the E-ticket ride on the global warming gravy train. But what about NASA's responsibility to the American taxpayers who foot the bill? I doubt that was ever a matter of concern.

There is something fundamentally wrong with an agency that both makes measurements and is involved in an activity that makes use of those measurements for its politically-based agenda. That is a conflict of interest not at all unlike the jurisprudence prohibition of an individual serving both as prosecutor and witness in the same criminal court trial.

The very-flawed, model-based assertions of "anthropogenic (human caused) global warming", alleged data-tampering, and fear-mongering by Al Gore, NASA, and others has, I allege, a far more sinister purpose: It provides a "scientific basis" for the largest international science-based scam ever perpetrated.

Numerous members of the United Nations' Intergovernmental Panel on Climate Change (IPCC) along with members of the American Geophysical Union, many of whom are supported by grants from NASA and NSF, have failed to tell the full truth about climate change. These scientists promote the claim that greenhouse gases, most especially anthropogenic carbon dioxide, are responsible for global warming. They remain silent about the consequences of the covert aerosol geoengineering (Figure 7) that has been taking place at least since at least the 1990s with growing scope and intensity and which about 2010 became a near-daily, near-global activity [9,10]. Failure to discuss this massive global anthropogenic phenomenon not only negates the validity of these scientists' assertions about climate change, but, I allege, makes those individuals, and their associated institutions, party to the biggest science-based scam ever perpetrated. And, as well, party to an activity many consider to be a crime against humanity and the environment [11].

Figure 7. Jet-sprayed geoengineering aerosol particulate trails across the February 4, 2017 sky in Soddy-Daisy, TN (USA). Courtesy of David Tulis. Millions of people have protested and expressed (well warranted) concern for their health and for the environment [12].

For 70 years the military interest in controlling the weather has been thoroughly documented [13]. Military experiments advanced from causing rain and snow to inhibiting rainfall by emplacing pollution particles into the atmosphere where clouds form.

Eventually, the atmosphere becomes too moisture-laden and torrential rains and storms result. The net effect of the now near-daily, near-global ongoing covert geoengineering activity of emplacing pollution particles into the atmosphere is to contribute to global warming and cause climate chaos. While some sunlight is reflected back into space by aerosols, the deliberately spread pollution particles also heat the atmosphere and impede heat loss from Earth [9]. The albedo of snow and ice is lowered by certain particulates when they fall to Earth.

Climate science that does not take into account the ongoing aerosol geoengineering activities can rightly be deemed immoral – by omission, lies are deliberately perpetuated. Such a dishonest effort to understand the climate is like trying to understand and predict ocean tides without considering the existence and gravitational influence of the Moon. NASA-funded climate scientists and IPCC scientists have been grossly remiss in ignoring the ongoing aerosol geoengineering. Not only are their climate-science results corrupted, but those scientists, and NASA officials who are unwilling to admit the evidence for ongoing climate engineering, demonstrate an absence of concern for human and environmental health.

Even without knowing the specific composition of the pollution particulates being sprayed into the air we breathe, some of the potentially adverse effects can be inferred from epidemiological studies of aerosol pollution in the same particle size range. As those studies have shown [14], pollution particles in the size range (PM2.5) are associated with Alzheimer's disease [15,16], lung cancer [17], risk for stroke [18], risk for cardiovascular disease [19], lung inflammation and diabetes [20], reduced renal function in older males [21], morbidity and premature mortality [22-24], decreased male fertility [25], low birth weight [26], onset of asthma [27], and increased hospital admissions [28].

Forensic investigation results are consistent with coal fly ash, the toxic-nightmare waste product of coal-burning utilities, likely being the main geoengineering aerosol in use world-wide [29-32]. Aerosolized coal fly ash poses risks to human health that include lung cancer [33], respiratory disease [34], and neurodegenerative disease [35]. The chemical components of coal fly ash present a range of toxic health hazards to humans and other biota, including poisoning the environment with chemically mobile aluminum [29] and mercury [36]. No one has the right to poison our planet, including the air we all breathe, and no one has the right to hide the health risks from the public. But NASA is all about deceiving the public about this matter.

The ongoing covert aerosol geoengineering program has been conducted without public disclosure, buttressed by deception [37-40]. In a 2005 document the U. S. Air Force lied, claiming that such an activity did not exist, deceiving the public that the particulate trails were harmless ice-crystal contrails [38]. Since 2005 and continuing to the present, NASA has been party to that deception and in one instance in a particularly deplorable manner: NASA instructs teachers to teach children to "count the contrails" [41] which in the great majority of instances are not ice-crystal contrails, but toxic particulate geoengineering trails that will contaminate those children and inevitably cause some of them to later suffer lung cancer [33], respiratory disease [34], neurodegenerative disease [35], and perhaps other maladies.

Spraying particulate pollution into the atmosphere where clouds form began as a military activity designed to weaponized weather. Then, sometime about 2010, that activity became internationalized, presumably by secret agreement, presumably connected to the United Nations. In an Eos interview [42], published April 18, 2018, former NASA Administrator (2009-2017) Charles Bolden stated: "We interact with more than 120 countries around the world...." The question to be answered is this: Did NASA through said interact[ions] sell out the health and well-being of millions of Americans to become party to the biggest science-based political scam [10] ever perpetrated?

Chapter 3. Nuclear Reactor Debacle

The planet Jupiter, one of the brightest objects in the night sky, has been known since ancient times. In 1610, Galileo Galilei pointed his custom made telescope at Jupiter and discovered four objects, which appeared to pass before Jupiter in a straight line, changing positions nightly. He correctly identified those objects as moons of Jupiter (Callisto, Europa, Ganymede and Io). Galileo, however, apparently did not observe the now well know atmospheric turbulent features that characterize Jupiter, especially its Great Red Spot, which some believe was first observed by Giovanni Domenico Cassini around 1665.

Consider the four giant planets of our Solar System (Figure 8). Note that three of the four planets exhibit pronounced turbulent features; Uranus is the exception.

Figure 8. Note the pronounced turbulence in the atmospheres of Jupiter, Saturn, and Neptune, but not conspicuous in the atmosphere of Uranus.

Changes have been observed in Jupiter's Great Red Spot. In the summer of 1878, Jupiter's Great Red Spot increased to a prominence never before recorded and, late in 1882, its prominence, darkness, and general visibility began declining so steadily that by 1890 astronomers thought that the Great Red Spot was doomed to extinction. Changes have been observed in other Jovian features, including the formation of a new lateral belt of atmospheric turbulence [43]. Dramatic appearance changes in atmospheric turbulence have recently been observed, not only in Jupiter, but in Saturn and Neptune as well.

Until the late 1960s, little documented thought seems to have been given as to what that atmospheric turbulence might mean regarding internal energy production. Before about 1969, scientists believed that planets do not produce energy internally, except for tiny amounts from the decay of a few radioactive elements; planets simply receive energy from the Sun, and then re-radiate it. Then, in the late 1960s, astronomers discovered that Jupiter radiates about twice as much energy as it receives from the Sun. The same was soon found to be true for Saturn and Neptune [44]. The major atmospheric turbulence observed in the giant planets appears to be driven by their internal energy sources. Jupiter, Saturn, and Neptune produce prodigious amounts of energy and display prominent turbulent atmospheric features. Uranus, on the other hand, radiates little, if any, internally generated energy and appears featureless in the main. The big question, of course, was what is the source of that energy? In 1990, this giant-planet internal energy production was described as "…one of the most interesting revelations of modern planetary science" [45]. For nearly two decades NASA-supported planetary scientists thought that they had considered and eliminated each possible candidate for Jupiter's internal energy production, declaring "by default" or "by elimination" that Jupiter's internally-generated energy is the consequence of gravitational collapse during planetary formation that began about 4½ billion years ago and, they assumed, continues into the present.

In 1978, David J. Stevenson [46], discussing Jupiter, stated, "The implied energy source ... is apparently gravitational in origin, since all other proposed sources (for example, radioactivity, accretion, thermonuclear fusion) fall short by at least two orders of magnitude…". Similarly, in 1990, William B. Hubbard [45] asserted, "Therefore, by elimination, only one process could be responsible for the luminosities of Jupiter, Saturn, and Neptune. Energy is liberated when mass in a gravitationally bound object sinks closer to the center of attraction ... potential energy becomes kinetic energy …."

Most of Jupiter, about 98%, is a mixture of hydrogen and helium. Both hydrogen and helium are excellent heat transport media. Moreover, Jupiter's turbulent atmosphere is indicative of active heat transfer. So, the idea that Jupiter is still collapsing and heat is still being given off after 4½ billion years did not seem logical to me, especially as Neptune, which is only about 5% of the mass of Jupiter, also radiates internally-produced energy.

In a particular area of science, when the pieces do not seem to fit together in logical, causally related ways, it is like red flags waving and red lights flashing, signals that something might be wrong with present understanding. For me, that was the case concerning the internally-produced energy in the giant gaseous planets. Was it possible that for two decades NASA-supported planetary scientists had overlooked a fundamentally different planetary-scale energy source? I pondered and pondered. Then one day, while shopping for groceries, the pieces suddenly fell into place. Like a bolt of lightning, I realized that Jupiter has all the ingredients for a naturally occurring nuclear fission reactor at its center.

Nuclear reactor – the mention of those two words might bring to mind names like Chernobyl or Three Mile Island or perhaps conjure images of complex mega-machines whose control rooms have more instrumentation than the cockpit of a 747 (Figure 9). The complex instrumentation belies the elegant simplicity that is the basis of operation of a nuclear reactor.

Figure 9. Inside a nuclear reactor control room, Fukushima 1. Courtesy of the Kawamoto Takuo.

At its heart a nuclear reactor is simply an accumulation of very heavy atoms, such as uranium, whose nuclei will split when hit with a neutron. When hit with a neutron, the very-heavy atomic nucleus breaks into two parts, a process called nuclear fission, liberating energy plus a few more neutrons. These just-freed neutrons can strike another nucleus causing it to fission; it can happen again and again and again, etc. as a chain reaction (Figure 10). This is a nuclear reactor. If too many neutrons come out of play, by escaping or by being absorbed, the chain reaction cannot be sustained.

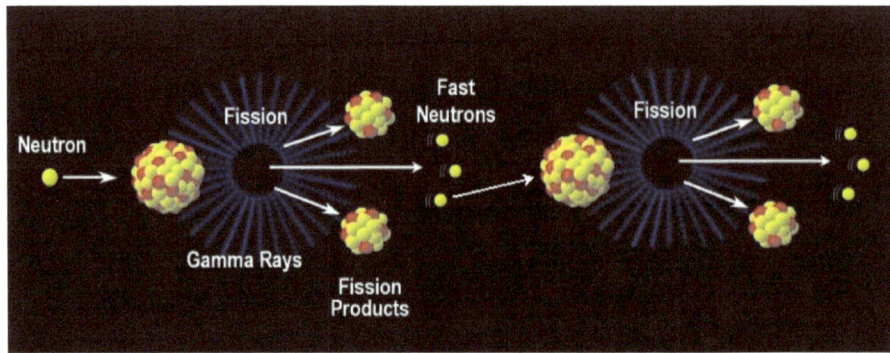

Figure 10. Schematic representation of the nuclear fission chain reaction (courtesy of Jaroslav Franta).

Enrico Fermi formulated nuclear reactor theory and in 1942 designed and built the world's first man-made nuclear reactor. In 1956, Paul K. Kuroda (Figure 11) used Fermi's nuclear reactor theory to show that nuclear fission chain reactions could have occurred in uranium deposits two billion years ago and earlier [47].

Figure 11. Paul Kazuo Kuroda (1917-2001) ca. 1957.

In 1972, French scientists discovered the intact remains of a natural nuclear fission reactor in a uranium mine at Oklo, in the

Republic of Gabon in Western Africa (Figure 12) [48]. The reactor had functioned almost two billion years ago just as Kuroda had predicted. Later, other fossil reactors were discovered in the region.

Figure 12. Vein of uranium ore in a mine at Oklo, Republic of Gabon that had functioned as a natural nuclear reactor 1.8 billion years ago (courtesy of Francois Gauthier-Lafaye).

At the time the remains of the Oklo natural nuclear fission reactor were discovered, I remember thinking that the discovery must have huge implications, but there were just too many pieces missing from the puzzle to progress further; it was like looking out into a very, very dense fog. Over the next two decades, without consciously realizing it, I began to fill in the missing pieces. Those pieces came together when I realized that Jupiter has all the ingredients for a naturally occurring nuclear fission reactor at its center.

Think about some of the considerations involved in demonstrating the feasibility of nuclear fission reactors as energy sources for the giant planets: (1) There must be a natural way for uranium to be concentrated; (2) There must be described a basis for natural planetary reactors to be able to operate in the present, and; (3) There must be a natural way for fission products, which are reactor poisons, to be able to be removed from the reactor.

(1) Concentrating Uranium: Uranium is a relatively low-abundance trace element. For a planetary-scale nuclear reactor to occur there must be a natural way for uranium to become concentrated. At the pressures that prevail inside of planets, particularly in the giant planets, density depends only on the element's atomic number (its number of protons) and the element's atomic mass (its number of protons plus its neutrons). Uranium, element number 92, being the heaviest element would be the densest element within the planet and would tend to concentrate by gravity at the planet's center, either directly or perhaps through a series of steps. A nuclear reactor is essentially an accumulation of reasonably pure uranium or a compound thereof. Paul K. Kuroda had already shown that under the appropriate conditions seams of uranium ore one meter thick could function as nuclear reactors. Kuroda's theoretical considerations [47] were verified by the discovery and investigations of intact fossil remains of nuclear fission reactors in a uranium mine at Oklo, in the Republic of Gabon in Western Africa [48-51].

(2) Fuel Breeding: Kuroda, like many other scientists, knew that at the present time the amount of U-235, compared to U-238, is too low in natural uranium for self-sustaining chain

reactions to commence; the U-238 absorbs too many neutrons. But, Kuroda realized that U-235 decays at a faster rate than U-238, so that, as he showed, about two billion years ago and earlier, when the relative amount of U-235 was greater, natural nuclear reactors were possible. So, under what conditions might a planetary-scale nuclear reactor be able to operate today? When Kuroda applied Fermi's nuclear theory, he assumed that the hydrogen in water would slow the neutrons and that fission would take place by those slow neutrons. Later, scientists studying the remains of fission fragments found in the reactor zones at Oklo discovered that yes, many of the fission events had in fact taken place by slow neutrons. But they also discovered evidence of fast neutron fission and of the production of plutonium. Some commercial nuclear reactors are breeders, changing U-238 into a more fissionable element, such as plutonium, by fast neutron bombardment. The observational evidence that natural nuclear fission reactors could be breeders formed a crucial piece of the logical ordering, for that meant that planetary-scale nuclear fission chain reactions, if initiated billions of years ago, could continue to the present through fast neutron fuel breeding reactions.

(3) Removing Fission Products: Another crucial matter for the concept of planetocentric nuclear reactors pertains to nuclear reactor survivability. A nuclear chain reaction requires the mass of the uranium to be sufficiently large so that not too many of the neutrons produced during fission escape and requires the uranium to be sufficiently pure so that not too many of the fission-produced neutrons are consumed in non-

fission activities. Typically, a nuclear reactor chain reaction splits uranium atoms into two fragments, called fission products. These fission products, if left in place, dilute the uranium, absorb neutrons, and will eventually shut down the chain reaction. I realized that for planetary-scale nuclear reactors there is a natural mechanism for removing fission products and for re-concentrating the uranium. At the pressures that prevail inside of planets, density depends only on the elements' atomic number and atomic mass. Because fission products are about half the atomic number and half the atomic mass of uranium, they will be less dense and will tend to migrate outward, while the uranium settles downward by gravity.

To demonstrate the feasibility of nuclear fission reactors for the giant planets, I did as Kuroda had done and applied Fermi's nuclear reactor theory. I described the background, basis, and results in a scientific paper entitled "Nuclear Fission Reactors as Energy Sources for the Giant Outer Planets", and in 1991 submitted it to the German journal *Naturwissenschaften* where in 1992 it was published [52]. This was a fundamentally new idea, an idea that planetary scientists had overlooked. This was a potential solution to "...one of the most interesting revelations of modern planetary science" [45]. This was the beginning of a whole new logical progression of understanding, like discovering a new path through the wilderness. As a consequence much, much more understanding would follow about the nature of planets and about the nature of the National

Aeronautics and Space Administration, NASA officials, and those to whom NASA provides financial support.

NASA formally and openly solicits research proposals from members of the scientific community, as NASA itself conducts few scientific investigations. On the surface, NASA's research proposal system might seem to be a sound way of bringing innovative new concepts into its space exploration program. So, when my *Naturwissenschaften* paper was accepted for publication [52], I submitted a research proposal to NASA's Planetary Geophysics Program. Paul K. Kuroda, who had predicted the natural reactors later discovered at Oklo, accepted my invitation to join in as a co-investigator. Kuroda, however, insisted that his efforts be pro bono as he "did not need the money".

The Universities Space Research Association, an association of major institutional recipients of NASA funding, operates the Lunar and Planetary Institute, which, at the time I submitted the proposal, operated the Lunar and Planetary Geoscience Review Panel (LPGRP). The LPGRP served NASA by soliciting secret reviews of submitted proposals, then in secret session reviewing the proposals, and ranking those so to make it easy for a NASA official to decide which to fund. The LPGRP, composed a group of principal investigators of NASA grants and contracts funded either through NASA's Planetary Geophysics Program or Planetary Geology Program, conducted the secret ranking of proposals submitted to one or the other of those same two NASA programs. In other words, my proposal was competing for the same limited pool of funds as proposals from the very institutions whose personnel served on the LPGRP. As disclosed by NASA, at the time, the chairman of the LPGRP, Torrance V. Johnson, was associated with NASA's Jet Propulsion Laboratory (JPL), which is operated by the California Institute of Technology (Caltech), and which consumed more than 40% of the budget of the Planetary Geophysics Program.

Whoa! What's wrong with this picture? Does anyone in authority at NASA understand common human behavior? NASA allowed individuals from NASA-funded institutions to be in charge of selecting reviewers, evaluating their reviews, and according a ranking to a proposal from a non-NASA-funded institution about a potential solution to "…one of the most interesting revelations of modern planetary science" that NASA-funded investigators had overlooked for twenty years!

Needless to say, my proposal was not funded. Normally, the LPGRP's ranking of proposals is kept secret, but through extraordinary efforts I learned from the U. S. Congress' General Accounting Office (since 2004 called Government Accountability Office) that on technical merit the LPGRP ranked my proposal lowest of the 120 proposals submitted to NASA's Planetary Geophysics Program. The lowest!

One might seriously question the integrity of that ranking, as I later independently performed all that I had proposed and much, much more, and the concept of planetary nuclear fission reactors has received quite thorough vetting in the international scientific community [53-59]. What can those institutions that receive funding from NASA do? Apologize? Make amends?

Admit incompetence or wrongdoing? No, of course not! That's not the NASA way. Their options:

- Pretend that the concept of planetary nuclear reactors never existed;
- Attempt to suppress scientific papers relating thereto;
- Make pejorative remarks to the press; and,
- Intimidate the scientific community with the implied threat of career-damaging consequences if anyone acknowledges that concept.

For two and a half decades, NASA-funded individuals have misled the scientific community and the public, effectively replacing scientific ethics with the political behavior of NASA's organizational culture. In the process, hand-in-hand with NASA officials, they have crippled NASA's ability to understand its own observations and contributed in a major way to the dumbing-down of American science and science education.

An article in the May 2, 1995 issue of *Eos, Transactions, American Geophysical Union*, entitled "Neptune's Nemesis", described observations of a new dark spot in the atmosphere of the planet Neptune. In addition to having a historical error, the article failed to represent to the geophysics community the significance of the observation with respect to possible on-going changes in the planetary driving-energy source. I responded with a brief, 500 word manuscript retort. In submitting the paper to Eos, I specifically requested that the manuscript not be sent to NASA's Jet Propulsion Laboratory (JPL) or to the California Institute of Technology (Caltech) that operates JPL for NASA because of a possible institutional conflict of interest.

Most publishers of scientific articles, including the American Geophysical Union (AGU), publisher of Eos, have policies clearly stating that editors should avoid real or perceived conflicts of interest. But in blatant contradiction to AGU policy on the avoidance of real or apparent conflict of interest, I was told by the managing editor, Laura Lawson, that Matthew Golombek of JPL would serve as section editor and that my only other option was to withdraw the article. Not surprisingly, Golombek demanded as a condition for publication that I remove all mention of the significance of the observations with respect to possible on-going changes in the planetary driving-energy source which included references to my published work on planetary nuclear fission reactors. There was no legitimate basis for such a demand. This was clearly unwarranted science suppression by an employee of NASA's JPL.

I appealed to the Eos Editor-in-Chief, AGU's Executive Director, A. F. Spilhaus, Jr. who, without ever providing substantive scientific criticism by competent referees, reiterated Golombek's demand that I remove all mention of the significance of the observations with respect to possible on-going changes in the planetary driving-energy source which included references to my published work on planetary nuclear fission reactors. To justify his demand, Spilhaus misrepresented the manuscript's content, falsely asserting that manuscript contains "original hypotheses regarding nuclear energy source" and that "Eos is a news and review journal, not a publication avenue for new scientific and technical ideas, which should be submitted for publication to a scientific journal." Even when advised that the manuscript contains no unpublished "original hypotheses", but is based upon published work [52-54], Spilhaus nevertheless persisted in his demand for censorship.

Next, I appealed to the AGU president, at the time Marcia Neugebauer, an employee of NASA's JPL. Not surprisingly, Neugebauer denied my appeal and "closed the case". To me, though, the case was not closed; suppression of scientific thought and scientific advances has no place in legitimate science under any circumstance and, especially, when perpetrated by individuals supported with taxpayers' hard earned money. Every two years, a new president takes the helm at the AGU. The next AGU president, following Neugebauer, was Sean C. Solomon, an individual with close financial ties to NASA. I brought my appeal to Solomon who denied it and, like his predecessor, closed the case. To me, though, the case was still not closed.

Then along came the next AGU president, John Knauss, an oceanographer with no ties to NASA. Knauss was appalled by what had happened with my article and assured me that a new submission would be handled properly and it was. So, in 1998 my new article, "Examining the Overlooked Implications of Natural Nuclear Reactors", was published by *Eos* [60]. It took three years to get published and it only got published because John Knauss that the integrity and the courage to stand up to dark side of the Force, spawned and encouraged by NASA, and paid for by taxpayers. And guess what? The world never stopped turning. NASA-supported institutions and their faculty/employees, however, never stopped deceiving the scientific community and the public and, presumably, will continue to do so as long as NASA continues to be given taxpayers' hard earned money by the United States Congress. Why? Because NASA's organizational culture is badly flawed, placing politics above science.

The purpose of science is to discover the true nature of Earth and Universe and to convey that knowledge truthfully to people everywhere. Science gives birth to technology that makes our lives easier and better. Science improves our health and enables us to see our world in ways never before envisioned, uplifting spirits and engendering optimism. Science is a continuing work-in-progress that involves replacing less-precise understanding with more-precise understanding.

Understanding is a distinctly individual process, not a committee activity; a logical process, not a democratic process. An individual ponders and through tedious efforts places seemingly unrelated observations into a logical sequence so that causal relationships become evident which can lead to new understanding and, perhaps, open the door to yet further discoveries.

When an important new idea emerges, it is the responsibility of the scientific community to attempt to confirm or to refute the concept. If the scientific community is unable to refute the new concept, then it should be cited in relevant subsequent scientific literature. Why? First, it is a matter of telling the truth, the full truth. Science is all about truth, not about lies, deception, or deceit. Second, someone somewhere, upon learning of the new concept, may realize a way to further advance the science. Those are the ethos of science. NASA, on the other hand, has fundamentally different ethos.

Eight years before the Space Shuttle Challenger disaster, NASA officials were aware of the potentially fatal "O-ring" design flaw, but did nothing that would have prevented disaster. NASA's "risk assessment" studies apparently led NASA officials to conclude that the estimated odds of a disaster did not warrant the political blow-back they perceived might result from delaying shuttle missions to fix the fatal problem. NASA officials rolled the dice, gambled with people's lives, to maintain their phony-baloney self-anointed, perception of NASA infallibility.

The same NASA ethos undergirds NASA's extreme departure from the long-standing principles of scientific integrity described above. Influential NASA-funded scientists willingly attempt to suppress publication of new advances and attempt to hide newly published, relevant concepts by systematically failing to cite or mention them. Even when made aware of that behavior, senior NASA officials elect to do nothing, content with continuing to deceive the scientific community, teachers, students and the public, while maintaining the impossible illusion of NASA's scientific credibility. So, how, by what mechanism, can independent-thinking, NASA-funded scientists become thus subverted?

Before World War II, there was very little government funding of science, but that changed because of war-time necessities. In 1951, the U.S. National Science Foundation (NSF) was established to provide support for post-World War II civilian scientific research. The process for administrating the government's science funding, invented in the early 1950s by NSF, has been adopted, essentially unchanged, by virtually all subsequent U.S. Government science-funding agencies, including NASA. There are several flaws to the science-funding methodology invented by NSF, but perhaps the most egregious is the concept of anonymous peer-review, where scientists review in secret their competitors' proposals for funding [1] [see Appendix]. Anonymous peer-review must have seemed like an administrative stroke of genius as the practice was almost universally adopted for reviews of scientific articles.

Under aegis of anonymity, an unscrupulous reviewer can cripple the career of his/her competitor with just a few negative or pejorative remarks while giving his/her own career and chance of being funded an undeserved boost. Scientists quickly learn to follow the in-crowd's NASA-funded research, and never acknowledge research that criticizes or contradicts the "consensus view" of the in-crowd; otherwise they fear that anonymous reviewers will castigate them in secret reviews of their proposals for funding or papers submitted for publication. If they challenge this procedure, they may as well bid farewell to a career in academic science.

I have published a host of scientific articles in world-class journals that bear directly on observations made by NASA [4,53-59,61-67]. To my knowledge, though, during the past two and a half decades, NASA-supported scientists have neither cited those articles nor mentioned the scientific content described therein. Even searches of the Science Citation Index®, which covers over 8,500 journals, failed to reveal any citations by NASA-funded authors. And, there are numerous instances where citation/mention should have been made. For example, in a book copyrighted in 2011 and 2006, the author, Linda T. Elkins-Tanton, a recipient of NASA support, describes Neptune's internal heat with the following statement: "The source of this unusual internal heat remains unexplained" [68]. Hello! I have explained the source of that internal heat in *Naturwissenschaften* [52] and in the *Proceedings of the Royal Society of London* [54] as being a natural nuclear fission reactor at Neptune's center (Figure 13). To date, no one has refuted in the scientific literature my published explanation or the calculations upon which it was based or the subsequent validation of those calculations [59].

Figure 13. Turbulence in the atmosphere of Neptune.

For twenty years before I demonstrated the feasibility of central planetary nuclear fission reactors, NASA-funded planetary scientists had promoted the idea that, even after some 4½ billion years, Jupiter is still collapsing and converting gravitational potential energy into kinetic energy. Now, twenty-five years after the first publication of my planetary nuclear fission papers [52-54], NASA-supported planetary scientists are, like before, promulgating the idea that the giant planets are still cooling, collapsing and releasing heat; themes and variations on the same old story. What is wrong now, though, is that they are not telling the full truth, never referencing planetary nuclear fission reactors. And this practice continues with Jupiter's moons.

There are many instances at the frontiers of science, at the interface of the unknown, where one cannot specify which of several possibilities might be correct or whether, perhaps, the correct possibility has not yet been envisioned. In such instances, ethical scientists cite the various suggested possibilities, including ideas other than their own that are published in world-class scientific journals. But, from my experience, that is not the NASA way. Decades ago, a coterie of influential NASA-funded scientists apparently decided:

- To exclude my scientific contributions;
- To attempt to suppress my publications, and;
- To systematically fail to cite my published scientific articles in instances where citation would have been appropriate.

Where was NASA's oversight? The failure of NASA officials to exercise scientific oversight has resulted in the trivialization of NASA's science and a severe and unnecessary loss to American prestige.

Figure 14. Jupiter's moon Ganymede which has an internally generated magnetic field.

The diameter of Ganymede (Figure 14), one of Jupiter's moons, is about 8% greater than that of planet Mercury. Like Mercury, Ganymede has an internally generated magnetic field which has apparently bewildered NASA-funded planetary scientists who state that the interior of the Ganymede would have cooled and solidified in the first 1-2 billion years making a convection-driven dynamo impossible [69,70]. But, of course, in a politically-convoluted manner, NASA-funded planetary scientists do not mention the possibility natural nuclear fission reactors. Perhaps, if they had, one of them might have realized that such a nuclear reactor is not only a heat source but potentially a self-regulated, convection-driven dynamo for magnetic field production. Of course, they didn't, but I did [57-59,65].

The objects in our Solar System, as in the Universe, are the consequence of the processes of their formation and the actions of subsequent processes. As a scientist, the best one can do is to try to understand those objects and the processes that brought them to their present states. But NASA-supported scientists must follow a different, unwritten rule, viz., ignore the work of others that might challenge the "approved consensus view" of the NASA in-crowd. I am a prime example: The advances I made have been systematically and thoroughly ignored, even though the underlying science is solid, relevant, and published in world-class scientific journals. How many capable scientists has this NASA behavior driven away? How many scientific advances have been lost? That is impossible to say. I am probably the exception. Rather than going away, I continued to make scientific discoveries at my own expense.

Those institutions and individuals who do in fact receive NASA funding for scientific investigations, however, must follow said unwritten NASA rule. Otherwise, in secret reviews, one or more of those NASA-funded individuals is apt to unwarrantedly berate proposals for funding or papers submitted for publication. Career-fear is pervasive in the world of NASA secret reviews. So, many remain silent and capitulate.

Consequently, others may never even have realized I made such relevant advances. That is the NASA-deception methodology and it has crippling NASA's science and science education. Many of the NASA-supported scientists are professors at prestigious American universities. These are the people who should be teaching students, in words and by their own example, that science is about telling the truth, the full truth. Instead, they are training generations of scientists never to contradict the "approved consensus view".

NASA has infected education with grants aimed at training science teachers or, perhaps better said, indoctrinating science teachers, including as mentioned before, indoctrinating teachers to deceive children about the aerial particulate spraying that poisons the very air they breathe. Science is all about telling the truth, the full truth, and about questioning extant ideas, observations, and theories. But how often do NASA-grant-recipient organizations teach teachers to question NASA's science? There is a widespread perception, real or imagined, that to question NASA's science will be grant-suicide. In this manner NASA takes an active part in dumbing-down American science education.

Figure 15. Jupiter's moon Io in near true color from the Galileo Mission. Note the volcanically scarred surface.

Jupiter's moon Io (Figure 15), with a diameter that is just 75% of Mercury's, is thought to have an internally generated magnetic field [70] and is currently the most volcanically active object in the Solar System [71,72] (Figure 16). At first, Io's volcanic eruptions were thought to consist of sulfur, which has a relatively low melting point, but later observations recorded much higher temperatures consistent with silicate volcanism [64, 65]. Io's surface heat flow has been estimated at more than

2.5 W/m² [73]. This is more than twice the heat calculated to arise from the gravitational push-and-pull (tidal interaction) with Jupiter and its other moons [74]. And, for NASA, that is hard to explain.

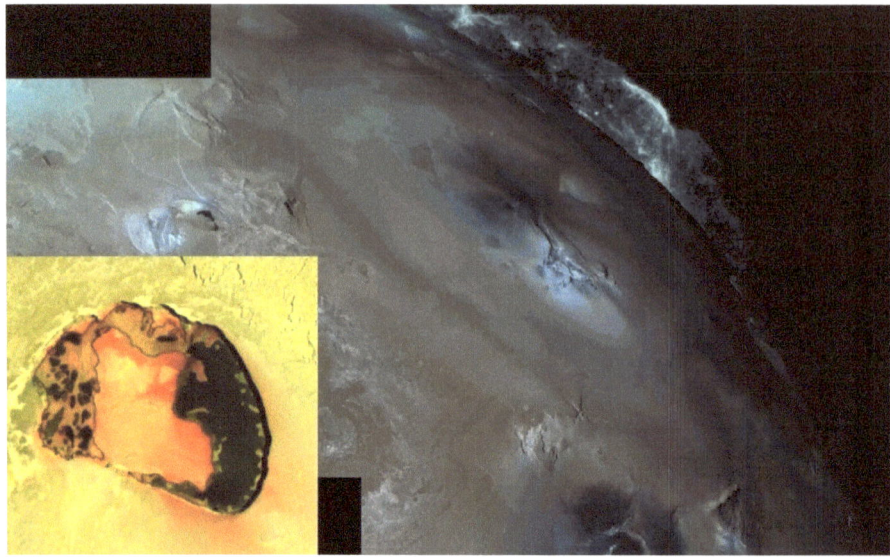

Figure 16. Volcanic eruption on Jupiter's moon Io. The volcanic plume rises 300 km above the surface in an umbrella-like shape. The plume fallout covers an area the size of Alaska. Inset shows volcanism at Tupan Patera on Io.

NASA-funded scientists are unwilling or unable to explain the great source of heat required for Io's silicate volcanism. Linda T. Elkins-Tanton wrote a series of books explaining the objects of the Solar System in terms laymen could understand. Sadly, though, she repeated the NASA-funded scientists' blatant misrepresentations that I describe in this book. Her own research relates to molten magma, so it is of interest to examine how she explains the energy shortfall related to Io's silicate volcanism.

Elkins-Tanton writes [69]: "Tidal stresses are not the final word …. Io's power output from volcanism is about 2.5 W/m². This value may be more than the energy contributed from Jupiter through tidal heating (some researchers contend that it is more than twice the energy that Jupiter contributes). It is about twice the magnitude of the heating provided by electromagnetic heating from the ion storm around Io, and it also exceeds any heat possible from radioactive decay. Io's heat output, in fact, is more than twice that of Earth's. Tidal heating is still accepted as the method for creating the heat required for the extravagant quantity of volcanic activity on Io."

The word "accepted" in the underlined last sentence begs the question: Accepted by whom? Accepted by the coterie of NASA-funded scientists that systematically fails to tell the full truth? NASA-funded scientists might pretend that it is their responsibility to accept or not accept science, but that is nonsense. The responsibility of NASA-funded scientists is to try to refute ideas they consider contradictory and, if they are unable to refute, then they should acknowledge those ideas. That is the basis of ethical, legitimate science. That is telling the truth, the full truth. As a practical consequence, Elkins-Tanton does a gross disservice to the general public by implying that "accepted" is an acceptable measure of scientific correctness. It is not.

So, what might Elkins-Tanton have written in place of the above underlined sentence? I suggest the following: It appears that Io contains an energy source not considered by NASA scientists, perhaps a Herndon-type nuclear fission reactor [52-59,65,75].

It is in the nature of science for a new concept to arise that explains what previously has been inexplicable. That happens again and again and in instances it engenders considerable jealousy and resentment. But the total acknowledgement-exclusion I have experienced from NASA-supported investigators has, I suggest, far more insidious implications.

Occasionally, a new concept burgeons into a whole new field of science and/or opens doors to new understanding in other areas of science. My demonstration of the feasibility of planetocentric nuclear fission reactors for the giant planets did both. Apparently, NASA-supported scientists were so involved in suppressing mention of my work that they gave no thought to the scientific consequences that would follow. Consequently, the concomitant discoveries and insights were left to me, rather than becoming rich new areas of investigation for the NASA-supported scientists to garner more research grants and contracts.

When I first considered the possibility of planetocentric nuclear fission reactors as energy sources for the giant outer planets, I thought that hydrogen would be necessary to slow the neutrons, much as water had slowed neutrons in the fossil reactors discovered in uranium mines near Oklo in the Republic of Gabon. Indeed, as I showed, the compound uranium hydride, UH_3, could serve both as fuel and as a moderator for slowing the neutrons. But then I started to wonder: What would happen if the heat produced were to tear apart the uranium hydride molecule and drive away the hydrogen? Would the nuclear chain reaction cease? Almost immediately the answer became apparent from current nuclear reactor technology; there would still be a nuclear reactor, a fast neutron reactor. Hydrogen was not necessary at all. This single bit of insight opened the door to the possibility of planetocentric nuclear fission reactors existing in planets like Earth that did not contain massive amounts of hydrogen. I already had reason to understand that a large percentage of the Earth's uranium resides in the core and, presumably, concentrated at its center and there is good reason to understand that a powerful energy source exists at or near the center of Earth. This is why.

Earth's layered structure, by the early 1930s, was asserted to consist of a fluid core, 3500km in radius, surrounded by a uniform 2900km thick solid rock shell, called the mantle, topped by a 10-50km thick crust. Analysis of records of a surprisingly large earthquake near New Zealand led Inge Lehmann (Figure 17) to discover the Earth's almost-Moon-sized inner core [76], correctly estimated to be 1200km in radius. Although decades later additional data led to refinement of Earth's interior composition, by 1940 this view of Earth's interior became the foundation for most textbook Earth science.

Figure 17. Inge Lehmann (1888-1993) and her discovery-diagram of Earth's inner core. For clarity, I traced the circles that bound the inner (red) and fluid (purple) core, and the region where earthquakes were claimed to be undetectable (blue). Ray #5 is reflected into that zone from the inner core she envisioned existed.

Four years after the discovery of the inner core by Lehmann [76], the idea of its composition being partially crystallized iron was birthed [77]. The rationale underlying the concept was this: In meteorites, such as ordinary chondrites, iron metal is invariably observed to be alloyed with nickel and the total relative abundance of elements heavier than nickel and iron is insufficient to form a body as massive as the inner core.

In the 1960s, two decades after the pronouncement that the inner core is partially crystallized iron metal [77], silicon was discovered in the metal of some enstatite chondrites [78]. Also, in the 1960s, a new mineral was discovered in enstatite chondrites, called nickel silicide or perryite, which consists of the elements nickel and silicon. Meteoritic nickel silicide occurs both as platelets that came out of solution (lamellar exsolutions) from silicon-bearing iron metal [79-83] and as more massive forms intimately associated with metal and iron sulfide in certain enstatite chondrites [81,84]. Nearly a decade later, I realized that if the Earth's core initially contained silicon, upon cooling the silicon would come out of solution by combining with nickel and precipitating as nickel silicide, leading to the formation of a mass virtually identical to the observed inner core mass [85]. I published the idea of the inner core being nickel silicide in the *Proceedings of the Royal Society of London* and received a complimentary letter from Inge Lehmann, discoverer of the inner core (Figure 18).

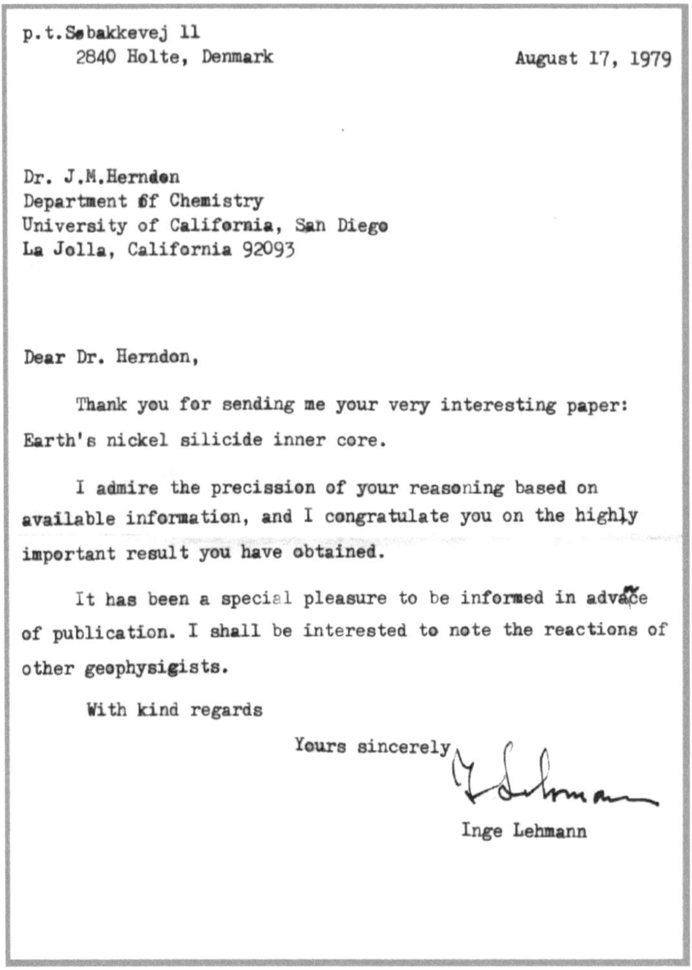

> p.t. Søbakkevej 11
>   2840 Holte, Denmark   August 17, 1979
>
>
> Dr. J.M. Herndon
> Department Of Chemistry
> University of California, San Diego
> La Jolla, California 92093
>
>
> Dear Dr. Herndon,
>
>   Thank you for sending me your very interesting paper: Earth's nickel silicide inner core.
>
>   I admire the precission of your reasoning based on available information, and I congratulate you on the highly important result you have obtained.
>
>   It has been a special pleasure to be informed in advace of publication. I shall be interested to note the reactions of other geophysigists.
>
>   With kind regards
>
>   Yours sincerely,
>   Inge Lehmann

Figure 18. Congratulatory letter from Inge Lehmann.

If the inner core were nickel silicide, I reasoned, then the interior of Earth must be like an enstatite chondrite, which is demonstrable [75,86,87] as only enstatite-chondrite-matter, not ordinary-chondrite-matter, has sufficient iron alloy to account for Earth's massive core. Moreover, I related the relative masses of enstatite chondrite minerals to the seismically-determined masses of the interior shells of the Earth (Figure 19), shown in Figure 20, which is based upon more recent data [4].

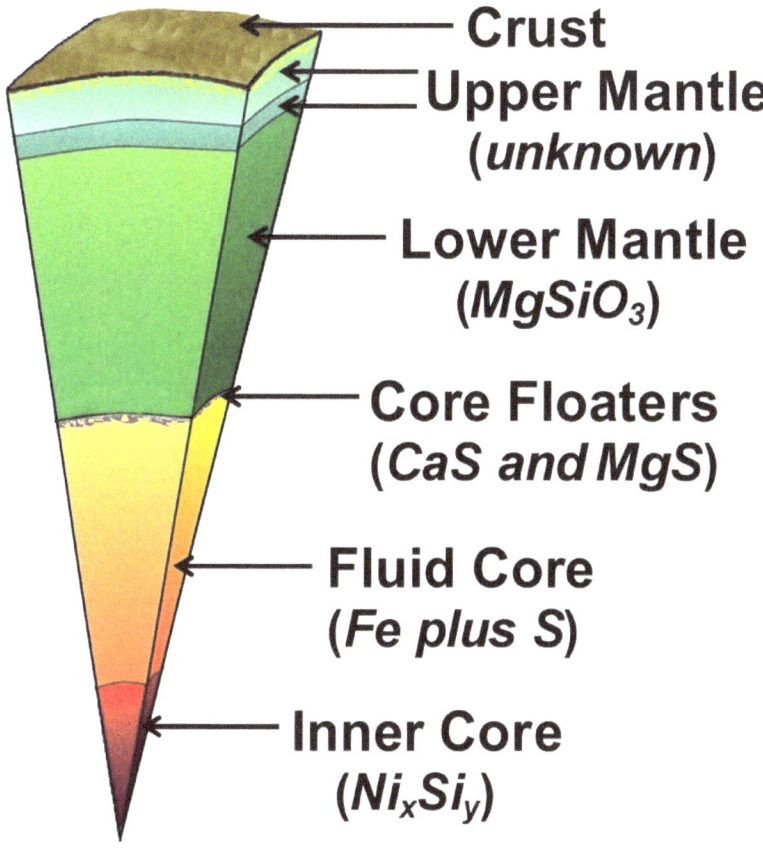

Figure 19. Chemical compositions of the major parts of the Earth, inferred from the Abee enstatite chondrite. The upper mantle, above the lower mantle, has seismically-resolved layers whose chemical compositions are yet known with certainty.

Fundamental mass ratio comparison between the endo-Earth (lower mantle plus core) and the Abee enstatite chondrite. Above a depth of 660 km seismic data indicate layers suggestive of veneer, possibly formed by the late addition of more oxidized chondrite and comet matter, whose compositions cannot be specified at this time.

Fundamental Earth Ratio	Earth Ratio Value	Abee Ratio Value
lower mantle mass to total core mass	1.49	1.43
inner core mass to total core mass	0.052	theoretical 0.052 if Ni_3Si 0.057 if Ni_2Si
inner core mass to lower mantle + total core mass	0.021	0.021
D" mass to total core mass	0.09	0.011*
ULVZ of D" CaS mass to total core mass	0.012	0.012*

* = avg. of Abee, Indarch, and Adhi-Kot enstatite chondrites
D" is the "seismically rough" region between the fluid core and lower mantle
ULVZ is the "Ultra Low Velocity Zone" of D"

Figure 20. Mass ratio relationships connecting seismically-determined parts of the Earth with mineralogically-determined parts of the Abee enstatite chondrite meteorite.

The Abee meteorite can be thought of as having two components: (1) The alloy portion consisting of metal and sulfides; and, (2) The silicate-rock portion. I discovered fundamental quantitative mass ratio relationships that connect the seismically-determined interior parts of the Earth with the mineralogically-determined parts of the Abee enstatite chondrite [75,86-88]. These mass ratio relationships, shown in Figure 20, demonstrate that below a depth of about 660 km, the Earth has the same state of oxidation as the Abee meteorite.

Uranium occurs in the portion of the Abee enstatite chondrite [89] that corresponds to the core of the Earth and, therefore, a major portion of Earth's uranium must occur in the core.

The Earth has a magnetic field originating at or near its center that requires a continuous supply of energy to replace the energy lost through its interactions with the matter of the Earth and with the solar wind. At the time, deep-Earth energy production was explained by ad hoc hypotheses based solely upon a series of arbitrary assumptions; namely, the inner core was assumed to be cooling and crystallizing, and heat produced by the assumed crystallization of iron metal was assumed to do useful work rather than simply slowing the rate of cooling. To me that explanation seemed contrived: i.e., it did not follow logically from the abundances of the elements and the properties of matter.

I did the background research, made the calculations by applying Fermi's nuclear reactor theory, and wrote the scientific article, entitled "Feasibility of a nuclear fission reactor at the center of the Earth as the energy source for the geomagnetic field", which was published in 1993 in the Japanese *Journal of Geomagnetism and Geoelectricity* [53]. In 1994 I published another work on planetary nuclear fission reactors in the *Proceedings of the Royal Society of London* [54] and in 1996 I published another article, entitled "Substructure of the Inner Core of the Earth", in the *Proceedings of the National Academy of Sciences (USA)* [55]. In that paper I described the rationale for the existence of a two-component substructure within the inner core, called the georeactor (Figure 21). The uranium sub-core of the georeactor is expected at the very center of Earth, surrounded by the sub-shell, composed of fission products and products of natural decay, such as lead. In that paper I pointed out that the sub-shell might be liquid or slurry and remarked: "The effect of nuclear activity on geomagnetic field production should not be discounted." These were embryonic ideas, seeds that would blossom into a new understanding of the origin of planetary magnetic fields [56-59,62,65,90].

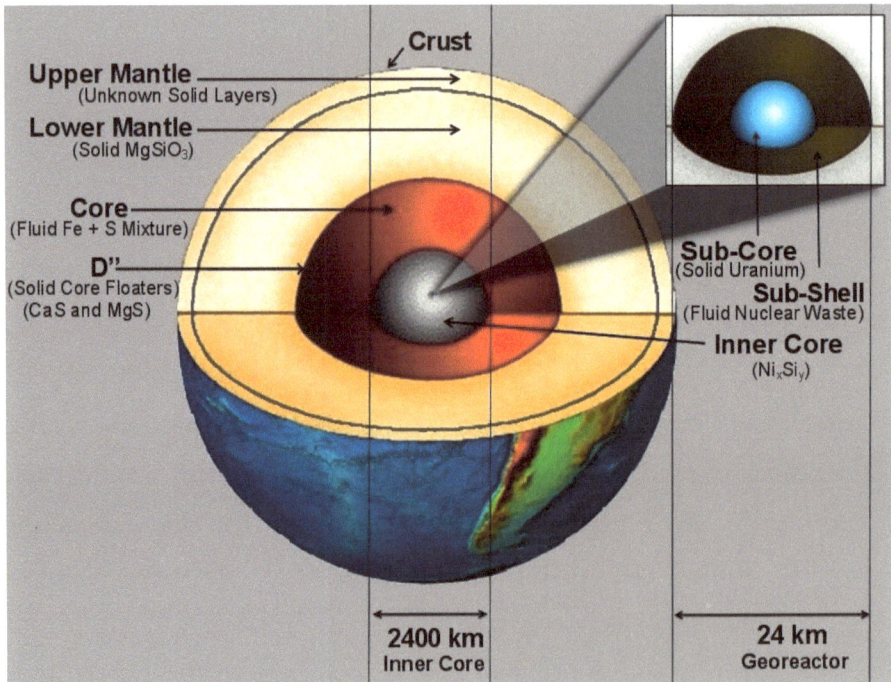

Figure 21. Schematic cut-away of Earth's interior and (inset) nuclear-fission georeactor at Earth's center.

The georeactor at the center is one ten-millionth the mass of Earth's fluid core. The georeactor's radioactive-waste sub-shell, I posit, is a liquid or slurry and is situated between the nuclear-fission heat source and inner-core heat sink, assuring stable convection, which is necessary for sustained geomagnetic field production by convection-driven dynamo action in the georeactor sub-shell.

When I first made the calculations demonstrating the feasibility of a natural nuclear fission reactor at the center of the Earth [52], I was thinking only of it as the energy source for powering the mechanism that produces the geomagnetic field. In 1939, Walter Maurice Elsasser began a series of scientific publications in which he proposed that the geomagnetic field is produced within the Earth's fluid core by an electric generator mechanism, also called a dynamo mechanism [91-93]. Elsasser proposed that convection currents in the Earth's electrically-conducting iron-alloy core, twisted by Earth's rotation, act like a self-sustaining dynamo, a magnetic amplifier, producing the geomagnetic field. For decades, Elsasser's dynamo-in-the-core was generally considered to be the only potentially viable means for producing the geomagnetic field and was generally cited without question. But, I discovered, there are serious problems, not with his idea of a convection-driven dynamo, but with its location, operant fluid, requisite "seed field", and energy source [57-59,62,65].

All electromagnetic generators involve motion. So, it was natural that Elsasser would consider a region within the Earth that was known to be liquid, the core, so that motion could occur, and he invoked the process of convection to keep the liquid in motion. For seventy years, since Elsasser's dynamo-in-the-core idea, the members of the scientific community have unanimously assumed that convection 'must' exist in the core.

Untold millions of dollars have been spent on modeling convection and its applications in the Earth's fluid core without anyone asking, "What's wrong with this picture?" But I did, and found that there are two serious problems that prevent sustained convection in the Earth's fluid core.

Because of the incumbent weight above, the bottom of the fluid core is about 16% denser (heavier) than the top of the core. The very small decrease in density, less than 1%, caused by thermal expansion is not enough to make a 'parcel' of bottom core material light enough to float to the top of the core. Moreover, the core-bottom cannot remain hotter than the top, as required for sustained convection, because the core is wrapped in a thick insulating rock-blanket; heat cannot readily escape as it does from the top of a pot of water on the stove-top. Further, the 'Rayleigh Number' [94], a calculation often used to justify Earth-core convection, I discovered, is inappropriate as it was derived for a thin film of liquid of uniform density. Without convection, there can be no convection-driven dynamo in the Earth's fluid core.

As I have described in the scientific literature [57-59,62,65] the nuclear fission georeactor has all the components for powering and generating the Earth's magnetic field by Elsasser's convection-driven dynamo mechanism. As illustrated in Figure 22, the georeactor has a sufficient energy source, its nuclear fission sub-core, and a region where sustained thermal convection is possible, its nuclear-waste sub-shell, where also a plethora of electrons are produced by radioactive beta-decay that may form the requisite magnetic "seed fields" for amplification by the convection-driven dynamo action.

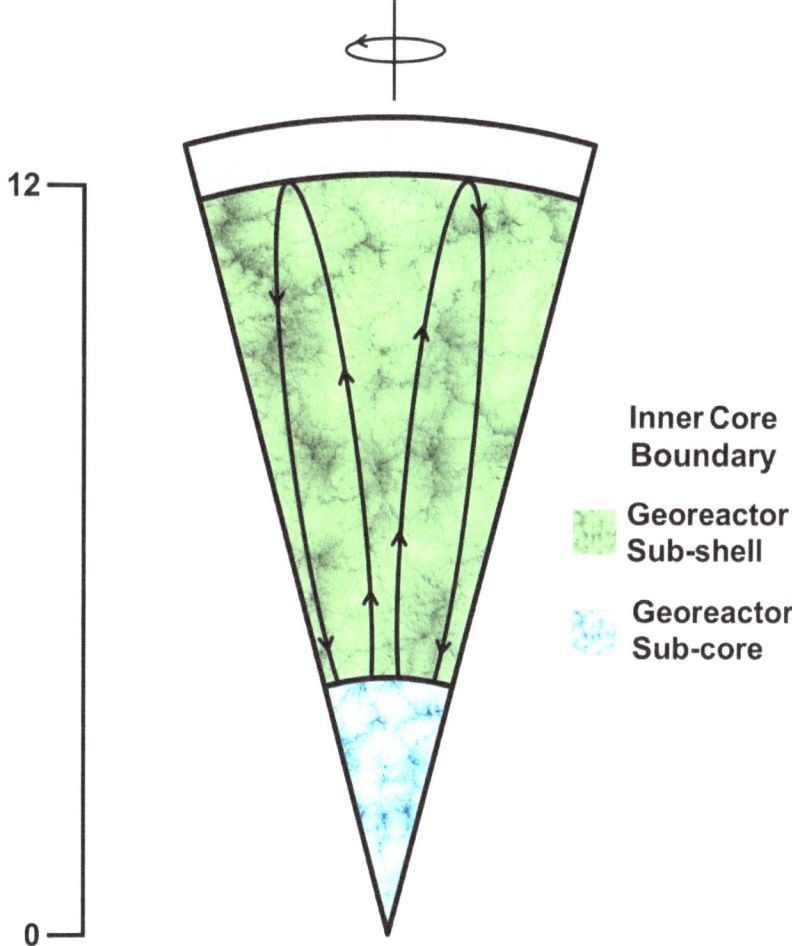

Figure 22. Schematic representation of the georeactor. Planetary rotation and fluid motions are indicated separately; their resultant or combined motion is not shown.

Stable convection is possible in the nuclear-waste sub-shell because the bottom (in contact with the nuclear sub-core) is hotter than the top where heat is removed by contact with the relatively more massive, thermally conducting inner core. The sub-core, surrounded by the nuclear waste sub-shell, is the region where stable convection is possible and where the geomagnetic field is generated by Elsasser's convection-driven dynamo mechanism. The scale at left is in kilometers.

NASA-supported scientists, in their haste to exclude my scientific contributions, excluded as well their own ability to extend my work to the science of planets. So, I extended it myself by demonstrating the feasibility of planetary-scale nuclear fission reactors in non-hydrogenous planets and large moons [57-59,62,65]. Before considering those instances, though, it is prudent to examine the evidence for a nuclear georeactor in the Earth.

For over thirty years, engineers and scientists at Oak Ridge National Laboratory have developed software to conduct computer simulations of the operation of different types of nuclear reactors. Their software has been validated by comparing calculation results with analyses of spent nuclear reactor fuel rods. A major advance was made when Daniel F. Hollenbach agreed to modify the software, making two changes, so as to accommodate the georeactor. One change allowed operating-times to extend to the age of the Earth and beyond. The other change made possible the removal of fission products. Why the latter change?

Deep inside planets and large moons, density depends only on atomic number and atomic mass. When a uranium nucleus splits, it produces two fission fragments, each having about half the atomic number and half of the atomic mass of the parent uranium. The fission fragments are thus less dense than uranium and migrate outward, forming a nuclear waste sub-shell surrounding the uranium sub-core.

The computer simulations made at Oak Ridge National Laboratory verified my previous calculations and expectations: The georeactor is indeed capable of functioning as a breeder-reactor over the entire age of the Earth at a power level within the estimated range for geomagnetic field production. Moreover, the Oak Ridge calculations provided data on the fission products that are unobtainable from calculations based upon Fermi's nuclear reactor theory. Helium fission product data led to the first substantial evidence for the existence of the georeactor [56,90].

Helium is a mixture of two components (isotopes), helium-3 and helium-4. In the 1960s, oceanographers observed that helium trapped in lava extruded from undersea volcanoes had a higher proportion of helium-3 than observed in atmospheric helium. For decades the origin of deep-Earth helium was a great mystery: The helium-4 was no surprise, because the alpha particles of natural radioactive decay become helium-4; the helium-3 was the mystery because there was no known deep-Earth production mechanism that could explain the amount of helium-3 observed. The explanation proffered for decades was that the helium-3 is primordial, trapped since Earth's formation and mixed with just the right amount of helium-4 to match observations. But then along came the Oak Ridge fission product results for helium.

When a uranium nucleus fissions, it generally splits into two roughly equal, large fragments. But once in every 10,000 fission events, the nucleus splits into three pieces, two large and one very small. Tritium, hydrogen-3, is a prominent very small fragment. Tritium is radioactive and decays to become helium-3. From the Oak Ridge calculations, the range of relative proportions of helium-3 to helium-4 have the same range as observed in oceanic lava (Figure 23). This is strong evidence that the georeactor exists and is the source of the deep-Earth helium.

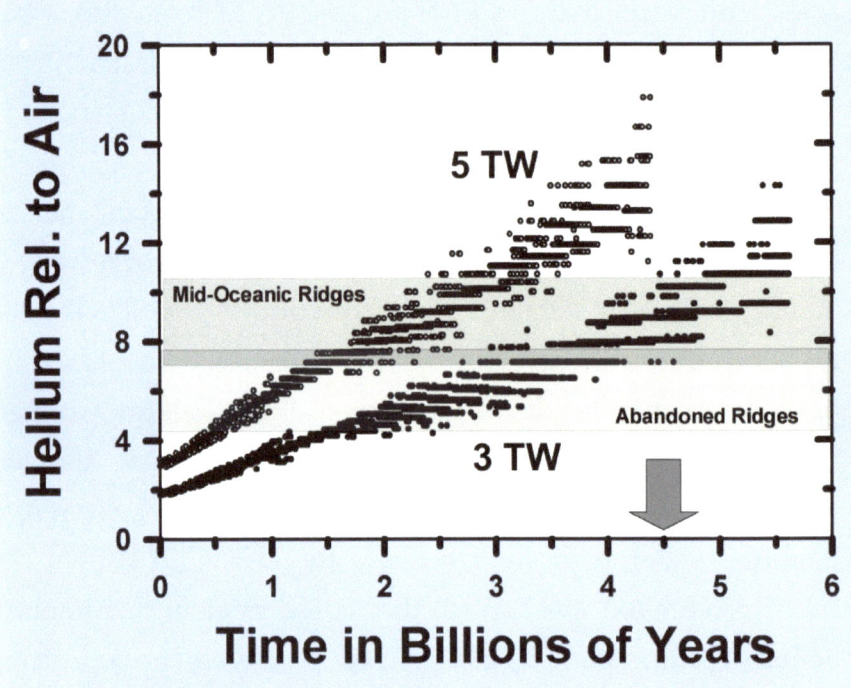

Figure 23. Fission product ratio of helium-3 to helium-4, relative to that of air for two Oak Ridge georeactor simulations.

In Figure 23, the band for measured values from mid-oceanic ridge basalts is indicated by the solid lines. The age of the Earth is marked by the arrow. Note the distribution of calculated values at 4½ billion years, the approximate age of the Earth. The increasing values are the consequence of uranium fuel burn-up. Icelandic deep-Earth basalts present helium ratio values ranging as high as 37 times the atmospheric value. TW stands for Terrawatts which is equal to 10^{12} watts.

As early as 1930, it seemed that energy mysteriously disappeared during the process of radioactive beta decay. The energy account sheet simply did not balance. To preserve the idea that energy is neither created nor destroyed, "invisible" particles were postulated to be the agents responsible for carrying energy away unseen. Finally, in 1956 these "invisible" antineutrinos were detected experimentally. Later, neutrinos from the Sun were detected as well as from a supernova. Antineutrinos were detected from nuclear fission reactors. It is not surprising then that R. S. Raghavan, a neutrino expert, after learning about the georeactor during a lunch-time seminar at Bell Laboratories, would author a paper, entitled "Detecting a Nuclear Fission Reactor at the Center of the Earth" [95]. Raghavan showed that antineutrinos resulting from nuclear fission products have a different energy spectrum than those resulting from the natural radioactive decay of uranium and thorium.

As early as the 1960s, there was discussion of antineutrinos being produced during the decay of radioactive elements in the Earth. In 1998, Raghavan was instrumental in demonstrating the feasibility of their detection. Now, Raghavan's paper on detecting a deep-Earth nuclear fission reactor stimulated intense interest worldwide, especially with groups in Italy, Japan and Russia. Russian scientists expressed well the importance: "Herndon's idea about georeactor located at the center of the Earth, if validated, will open a new era in planetary physics" [96].

Antineutrinos can literally fly through the Earth interacting very, very little, if at all, with the matter of Earth, which is the reason they are difficult to detect. Antineutrino detection requires large, extremely sensitive detectors operated for long periods of time and buried deep underground to shield from cosmic rays. To date, detectors Kamioka, Japan and Gran Sasso, Italy have detected antineutrinos coming from within the Earth. But, after years of data-taking, the first measurements have been made. From the total energy output of uranium and thorium, estimated from deep-Earth antineutrino measurements, an upper limit on the georeactor nuclear fission contribution was determined to be either 26% (Kamioka, Japan) [97] or 15% (Gran Sasso, Italy) [98].

The total georeactor contribution may be somewhat greater, though, as some georeactor energy comes from natural decay as well as from nuclear fission.

Following my 1992 publication of the feasibility of planetocentric nuclear fission reactors as energy sources for the giant outer planets, I extended the concept to non-hydrogenous planets, including Earth. The idea that a nuclear fission georeactor exists at the center of Earth and at the centers of other planets and large moons may, as noted by Russian scientists [96], "open a new era in planetary physics" for several reasons: A planetocentric nuclear fission reactor provides a previously unanticipated heat source for geodynamic activity as well as a previously unanticipated mechanism for magnetic field generation. Moreover, the nuclear reactor energy source is at the planet's center where it needs to be and where previous ideas are deficient.

The tiny planet Mercury (Figure 24), with a mass of about 5% that of Earth, is an excellent example of planetary circumstances that might be better explained by considering a central nuclear fission reactor. Mercury is also an excellent example of NASA-supported scientists systematically refusing to make any reference to my planetocentric nuclear reactor concept even though they make references to a variety of ideas that are on considerably less firm scientific footing.

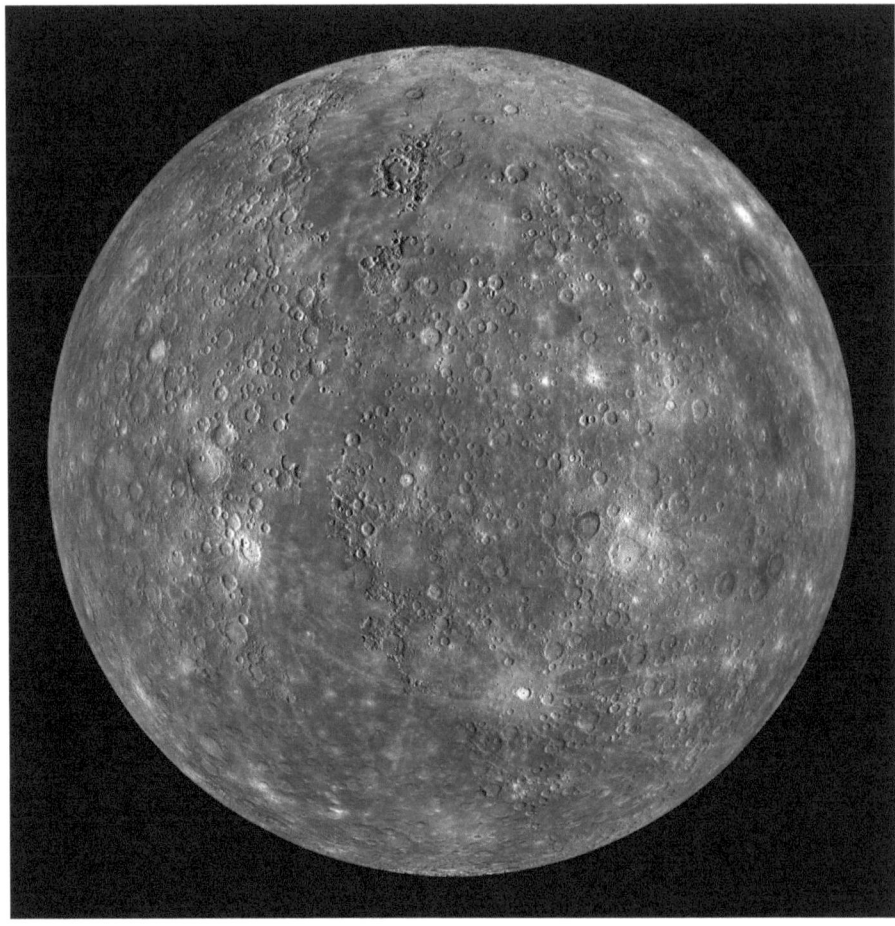

Figure 24. Planet Mercury, composite image from NASA's MESSENGER mission.

There has long been conflict related to Mercury's heat source and presumptive magnetic field generation. In 1974-75, the Mariner 10 spacecraft flew by Mercury three times and discovered that Mercury has a global magnetic field. The nature of that magnetic field and its energy source has been the subject of much discussion by NASA-supported scientists.

Before my concept of planetocentric nuclear fission reactors, in 1976 Sean C. Solomon wrote: "A convective dynamo mechanism for Mercury's magnetic field is in apparent conflict with cosmochemical models that do not predict a substantial source of heat, most probably radiogenic, in Mercury's core. Without such a heat source, the core would solidify within about 1 b.y. [billion year] after core infall, producing an unacceptably large contraction in Mercury's radius" [99].

For twenty years and continuing in the present, NASA-funded scientists, including Sean C. Solomon, the Principal Investigator for NASA's MESSENGER Mission to Mercury, have systematically and totally failed to mention or cite my planetary nuclear fission reactor scientific publications [52-58,75,90] when my concept could potentially explain the existence of an unanticipated, substantial heat source at Mercury's center as well as a unique, unanticipated mechanism for the generation of Mercury's magnetic field. Instead, NASA-funded scientists have published a plethora of core-dynamo models based upon arbitrary assumptions [100,101] and even published a thermoelectric model that requires even further assumptions [102]. But no Herndon references!

Science is about telling the truth, the whole truth, but what passes at NASA is a politically distorted version of the truth, which is not truth at all and which is not good science either. Distortions and deceptions extend throughout NASA's widely promulgated but very flawed 'understanding' of the Solar System. And it is paid for with taxpayer-provided funds. What's wrong with this picture?

Chapter 4. NASA's Faulty Storyline

Momentary streaks of light flashing across the night sky, called "shooting stars", are produced when meteors from outer space burn up as they come crashing to Earth. Occasionally, a meteor will survive the fiery trip through the atmosphere, land, and be recovered – then it is called a meteorite. Two centuries of investigations of meteorites by chemists, physicists, metallurgists, mineralogists and petrologists have led to a vast amount of data on hundreds of meteorites that form diverse groups, which are classified in different ways. Generally, meteorites are the remains, the bits and pieces, the evidence-in-hand of events and processes that occurred during the time the Solar System formed. From meteorites that land on Earth, one can obtain fundamental information about the nature of processes operant during the origin of the Solar System.

Consider the planets of the Solar System (Figure 25). The four inner planets are "rocky"; the four outer are "gas giants". Yet, there is good reason to believe that all planets formed from primordial matter of the same composition, matter like that in the outer part of the Sun. In the photosphere of the Sun, there are more of the "gaseous" elements than "rocky" elements. In fact, if our planet were put back together with its original complement of primordial gas, Earth would be about 300 times as massive, almost identical to Jupiter. So, what happened to the inner planets' gases?

Figure 25. Upper image: Comparison of the relative sizes of the planets of our Solar System [Mercury, Venus, Earth, Mars, Jupiter, Saturn, Uranus, and Neptune]. Note that the distances between these objects in the upper image are not to scale. Lower image: The distance of each planet from the Sun, expressed in Astronomical Units (AU), the distance between Earth and Sun.

Since the first hypothesis about the origin of the Sun and the planets was advanced in the latter half of the 18th Century by Immanuel Kant and modified later by Pierre-Simon de Laplace, various ideas have been put forward. Generally, concepts of planetary formation fall into one of two categories that involve either condensation at very high-pressures or at very low-pressures.

Beginning in the 1940s, the idea of condensation from within a giant gaseous protoplanet, with pressures of hundreds to thousands of atmospheres was discussed [103,104]. But then in

1963, A. G. W. Cameron [105] made a model based upon the assumption that primordial matter condensed from a hot gas of solar composition at a pressure of about one ten-thousandth of an atmosphere; the condensed dust then gathered into progressively larger particles, then into rocks, ultimately becoming planetesimals, which finally gathered to become planets. This modern version of the circa 1900 Chamberlain-Moulton "planetesimal hypothesis" became part of the NASA storyline often referred to as the "standard model of solar system formation". For the terrestrial planets, such a model also leads to the further assumption of a "magma ocean", whole-planet melting so that iron would drain to the planet's center forming the core [106].

Were it not for meteorites, debate about the nature of planetary formation might forever remain an esoteric theoretical endeavor. But with application of sound scientific methods, logical reasoning, and a strict adherence to truth, one can deduce, from the wide variety of diverse types of meteorites, the nature of processes and conditions that led to their formation.

Elemental abundances, expressed as ratios, measured in the laboratory of one group of meteorites, called chondrites, are quite similar to corresponding rock-forming element ratios determined spectrometrically in the outer part of the Sun. Primarily for this reason, elemental abundance ratios in the photosphere of the Sun are typically taken to represent the composition of the primordial matter from which planets and meteorites formed. So the fundamental challenge is to understand the conditions and the processes that transformed primordial "solar" matter into matter characteristic of chondrites, a trivial exercise as chondrites consist of three major groups, enstatite, carbonaceous, and ordinary, differing significantly in oxygen content and in other ways.

More than a decade before NASA was established, influential members of the geoscience community (wrongly) decided that our planet resembles an ordinary chondrite, the most common type of chondrite that falls to Earth. Their rationale was that carbonaceous chondrites contained no free iron metal, and therefore could not form a metallic core, and enstatite chondrites were very rare and contained minerals not found on the Earth's surface. This mistaken thought-belief spilled over into well-funded NASA-supported work and led to much confusion.

During the late 1960s and early 1970s, the so-called "equilibrium condensation" model [107] was developed by NASA-funded scientists to augment Cameron's model [105]. That condensation model was based upon the assumption that the mineral assemblage characteristic of ordinary chondrite meteorites formed as condensate from a gas of solar composition at pressures of about one ten-thousandth of an atmosphere. This condensate, it was assumed, would ultimately become the planets by the "planetesimal hypothesis."

The problem with the equilibrium condensation model is that a large part of the iron occurs as iron-oxide, FeO, incorporated in crystalline silicates, such as the mineral olivine [108]. In 1977, Hans Suess and I showed that the FeO content of those silicates is thermodynamically inconsistent with condensation from a gas of solar composition, and is more consistent with condensation from a gas that is depleted in hydrogen by a factor of a thousand relative to primordial matter [109]. In 1978, I showed that the gas must also have been depleted in oxygen by a factor of "several", implying a condensation process involving the re-evaporation of condensed matter after separation from the primordial gases. The latter paper was published in the Proceedings of the Royal Society of London with the title "Re-evaporation of condensed matter during the formation of the Solar System" [110].

NASA supports the making of models based upon arbitrary assumptions and based upon other models that are based upon arbitrary assumptions. Highly paid officials at NASA Headquarters provide little or no scientific insight or oversight. Moreover, they oversee a system that allows insiders to exclude competitors by means of secret reviews. Not surprisingly, science-nonsense readily becomes part of the "accepted" NASA story and, even when shown to be wrong, nevertheless remains a part of that story. One case in point: Condensation from a gas of solar composition at a pressure of about one ten-thousandth of an atmosphere. Recently published articles still quote "condensation temperatures" from those calculations and, far more seriously, few, if any, seem to understand or to acknowledge understanding the relevant model-flaw that I revealed [111].

Thermodynamic considerations demonstrate that, in a gas of solar composition at a pressure of about one ten-thousandth of an atmosphere, condensation is expected to progress at relatively low temperatures in a range of fairly oxidizing conditions. At low temperatures, all of the major elements in the condensate may be expected to be oxidized because of the great abundance of oxygen in solar matter relative to the other major condensable elements [112]. Beyond these generalizations, in this low-pressure regime, precise theoretical predictions of specific condensate compounds and condensation temperatures may be limited by kinetic nucleation dynamics and by gas-grain temperature differences arising because of the different mechanisms by which gases and condensate lose heat. The hydrous carbonaceous chondrite meteorites, such as Orgueil, have the state of oxidation and mineral components with characteristics similar to those expected to form from a condensate from solar matter at low pressures. Essentially all of the major elements in these few chondrites are oxidized, including sulfur.

The idea of planetary formation from a diffuse solar nebula, with hydrogen pressures of about one ten-thousandth of an atmosphere, envisioned that dust would condense at fairly low temperatures, and then would gather into progressively larger grains, and become rocks, then planetesimals, and ultimately planets. In the main, that picturesque idea leads to the serious contradiction of the terrestrial planets having insufficiently massive cores, because the condensate would be far too oxidized for a high proportion of iron metal to exist. In other words, the 'standard model of solar system formation' is wrong. But that flawed model continues to be touted by NASA-funded scientists. Why? Perhaps because of the perception, real or imagined, that one will lose NASA funding if one challenges others' NASA-funded work.

As evidenced by Orgueil and similar meteorites, such low-temperature, low-pressure condensation did in fact occur, perhaps only in the evolution of matter in the outer regions of the Solar System or in interstellar space [62], and thus may contribute to terrestrial planet formation only as a component of late-addition planetary veneer.

In 1987 Fegley and Palme [113,114] 're-discovered' what Suess and I had previously published [109,110] about loss of hydrogen and re-evaporation, but failed to cite our prior discovery. In 1995, Fegley's wife, Lodders, 're-discovered' [115] that the interior of Earth resembles an enstatite chondrite without citing my prior discovery [53,85,86,116] which is on substantially firmer footing, as I related the relative masses of enstatite chondrite minerals to the seismically-determined masses of the interior shells of the Earth (Figures 19 and 20).

The great majority of meteorites observed landing on Earth are ordinary chondrites. In the 1940s, when geoscientists collectively decided that Earth resembles an ordinary chondrite meteorite, they ignored enstatite chondrites, which are rarely observed landing. The enstatite chondrites are not only rare, but the minerals they contain formed under highly reducing conditions that seriously limited their oxygen content: Their origin was not at all understood. In 1975, the idea was advanced that for some unknown reason in the region of space where enstatite chondrite material formed, unlike as in primordial "solar" matter, carbon was somehow more abundant than oxygen [117]. In 1976, Hans E. Suess and I discovered from thermodynamic considerations a possible explanation for oxygen-starved enstatite-chondrite-like-matter arising as a consequence of condensation from solar matter at high-pressures [118].

The principle is simple: In a gas of solar composition, where hydrogen is a thousand times more abundant than iron, the availability of oxygen, expressed as oxygen fugacity [119], is governed by the pressure independent gas-phase reaction,

$$H_2 + \tfrac{1}{2}O_2 = H_2O$$

Condensation, on the other hand, is a direct function of pressure. At high pressures, substances condense at high temperatures where the above reaction causes oxygen to be very limited.

In 1944, the German physicist/chemist, Arnold Eucken (Figure 26), on the basis of thermodynamic considerations, suggested core-formation in the Earth as a consequence of successive condensation from solar matter, on the basis of volatility, from the central region of a hot, gaseous protoplanet with molten iron metal first raining out at the center forming the core before the mantle had fully condensed [103] obviating the assumption of whole-Earth melting and the formation of a magma ocean.

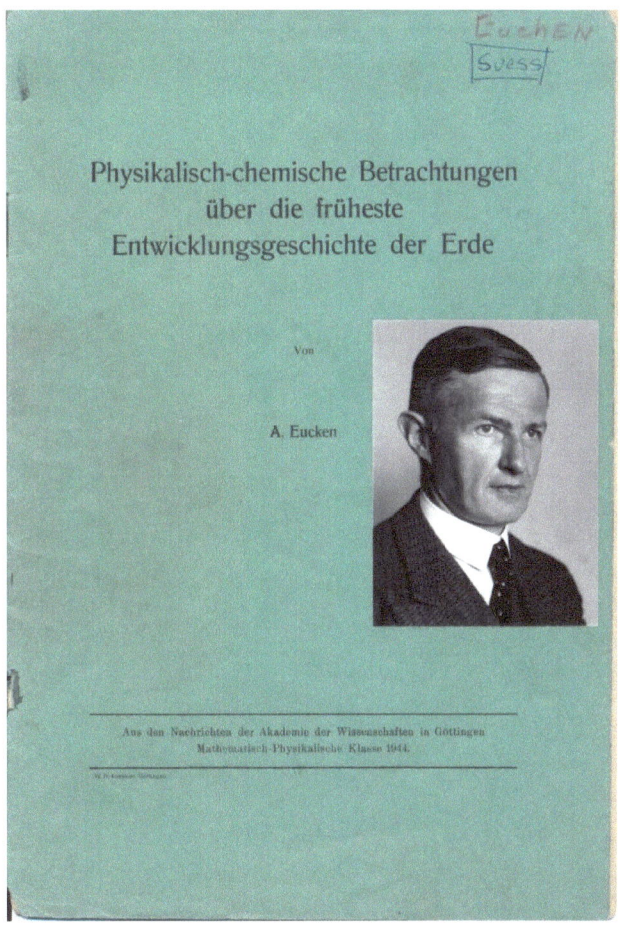

Figure 26. Re-print cover of the 1944 seminal scientific work of Arnold Eucken (1884-1950) on Earth formation with his photograph inset.

This is the connection: The condensate from within a giant gaseous protoplanet resembles an enstatite chondrite [62,118] and the interior of Earth, below 660 km, resembles an enstatite chondrite (Figure 20). The calculations of Herndon and Suess [118] complement those of Eucken [103]. Thus, one may reasonably conclude that the Earth formed by raining out from within a giant gaseous protoplanet. Complete condensation then led to a gas giant planet similar in mass to Jupiter. So, what process in nature could remove the giant gaseous envelopes from the inner four planets, including removing 300 Earth-masses of gas from our own planet?

In the early stages of star formation, when a star like our Sun begins to ignite, a brief period of violent activity occurs. This is the so-called T-Tauri phase; it is characterized by grand eruptions and super-intense 'solar-wind'. The Hubble Space Telescope captured images of an erupting T-Tauri phase binary star over an interval of five years (Figure 27). In five years, the leading edge of the eruption plume moved outward a distance equal to 130 times the distance from the Earth to the Sun. A similar T-Tauri outburst by our young Sun, I posit, stripped the primordial gases and ices from the inner four planets.

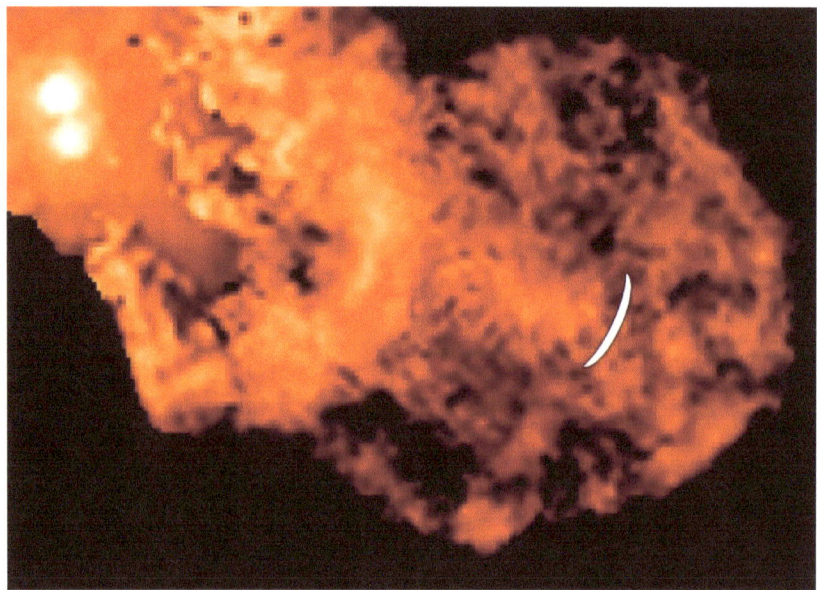

Figure 27. Outburst from a T-Tauri phase binary star in 2000. Hubble Space Telescope image of XZ-Tauri (2000).

In Figure 27, the white crescent label shows the position of the leading edge of that plume in 1995, indicating a leading-edge advance of 130 AU in five years. T-Tauri eruptions are observed in newly formed stars. Such eruptions in the pre-Hadean, I submit, stripped the primordial gases from the inner four planets of our Solar System.

So how do NASA-funded scientists explain why the terrestrial planets differ from the gas giant planets? The prevailing NASA "accepted" view is that, during the formation of the Solar System, the Sun had already ignited and its heat produced a temperature gradient throughout the plane of the "solar nebula". At a distance from the Sun of about 3½ AU (1 AU = the distance from Sun to Earth), a hypothetical "frost line" was calculated (Figure 28). Inside the hypothetical frost line the temperature was assumed too hot for ices to form so that only the rocky stuff of the terrestrial planets could condense. Whereas outside the hypothetical frost line, temperatures were sufficiently cold that ices and gases could condense and form the giant planets.

Figure 28. NASA artist's representation of the solar nebula with added green ellipse to show thehypothetical "frost line".

NASA-funded scientists have ignored my published work, which leads to an entirely different vision of planetary formation, with Earth's early formation as a Jupiter-like gas giant. As a result, NASA finds itself with an inexplicable conundrum.

The idea of giant planet formation at about 5 AU, well beyond the "frost line", was thoroughly entrenched in the NASA "accepted" storyline when gas giant exoplanets were discovered in other planetary systems as close to their star, or closer, than Earth is to the Sun. So, how were the close-to-star gas giant exoplanets explained? Bingo! Astrophysicist-modelers invented "planet migration" where gas giants were assumed to have formed at about 5 AU from their star and then were assumed to have migrated to their close-to-star locations.

In 2006, I attempted to publish a short paper in *Astrophysical Journal Letters*, entitled "Evidence Contrary to the Existing Exo-Planet Migration Concept". Usually before a paper is sent out for review, that journal requires the author to sign a document transferring copyright to the American Astronomical Society. In that instance, I was never asked to sign the copyright transfer document as the procedure requires. Instead my submission was reviewed in secret and promptly rejected. Strange? I did, however, post the preprint [120].

Five and a half years later, NASA's website listed four "Big Questions" one of which stated: "Many of the other solar systems have massive Jupiter like planets close to their Sun, closer even than Mercury. Many scientists now believe that these gas giants could not have formed there. Rather, they must have began (sic) out where our Jupiter is, and moved inwards, scattering the smaller planets with their powerful gravity as they went. Why is it that our Jupiter and Saturn did not migrate inward?..."

Such convoluted logic from America's supposedly premiere space agency: If astrophysicists do not understand how Mother Nature did something, such as form close-to-star gas giant planets, then obviously Mother Nature could not do it! Close-to-star gas giant exo-planets did not necessarily migrate inward; planet migration is just an invention to explain close-to-star exo-planets by NASA's badly flawed solar system formation models. Instead of questioning the validity of their badly flawed models, and perhaps learning from the flaws, NASA-funded investigators seek to add another layer of flaws.

There is a different explanation as to why the terrestrial planets have no primordial gases: T-Tauri outburst by our young Sun, similar to the one shown in Figure 27, I posit, stripped the primordial gases and ices from the inner four planets.

Chapter 5. Science NASA Missed

One of the grandest aspirations of NASA's planetary exploration, it might seem, would be to better understand our own planet's origin, evolution, and behavior. But NASA officials and NASA-funded investigators systematically excluded themselves from doing that. How so? By excluding my advances, and never citing my published contributions, NASA officials and NASA-funded investigators failed to join in my logical step-by-step progression of understanding. Instead, they became evermore lost in a labyrinth of model-nonsense, all at taxpayer expense.

I have described a new indivisible geoscience paradigm that begins with and is the consequence of our planet's early formation as a Jupiter-like gas giant [59,62,65] and which permits deduction of: (1) Earth's internal composition and highly-reduced oxidation state [75,87]; (2) Core formation without whole-planet melting [61,62]; (3) Powerful new internal energy sources, protoplanetary energy of compression and georeactor nuclear fission energy [59,62]; (4) Mechanism for heat emplacement at the base of the crust [63]; (5) Georeactor geomagnetic field generation [57,59]; (6) Decompression-driven geodynamics that accounts for the myriad of observations attributed to plate tectonics without requiring physically-impossible mantle convection [61]; (7) Fold-mountain formation that does not necessarily require plate collision [64], and; (8)

New understanding of the origin of fjords, lochs, and submarine canyons [66].

I have presented a new understanding of the origin of petroleum and natural gas deposits [121] and pointed out that supercontinent cycles [67], like planet-migration [120] are fictitious constructs, attempts to describe.

Briefly, the weight of 300 Earth-masses of primordial gases gravitationally compressed the non-gases: the original rock-plus-alloy kernel that became Earth to some 64% of its present radius, sufficient compression for solid continental-rock crust to cover the entire planet, as envisioned by Hilgenberg [122] in 1933. As the Sun ignited, its violent T-Tauri outbursts stripped Earth of its Jupiter-like gas envelope, marking the beginning of the Hadean eon (Figure 29).

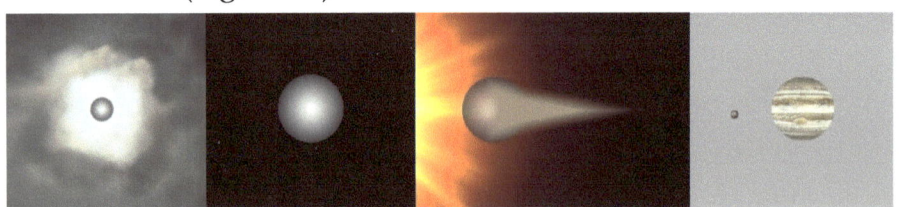

Figure 29. Whole-Earth decompression dynamics formation of Earth.

In Figure 26, from left to right, same scale: 1) Earth condensed at the center of its giant gaseous protoplanet; 2) Earth, a fully condensed as a gas-giant; 3) Earth's primordial gases stripped away by the Sun's T-Tauri super-intense solar-wind; 4) Earth at the onset of the Hadean eon, compressed to 64% of present diameter, shown by comparison with Jupiter.

Later, when internal pressures built up, sufficient to crack the rigid crust, the Earth began to decompress by expansion (Figure 30). Gravitational energy of compression that had been stored during Earth's gas-giant stage is the primary power-source for Earth dynamics.

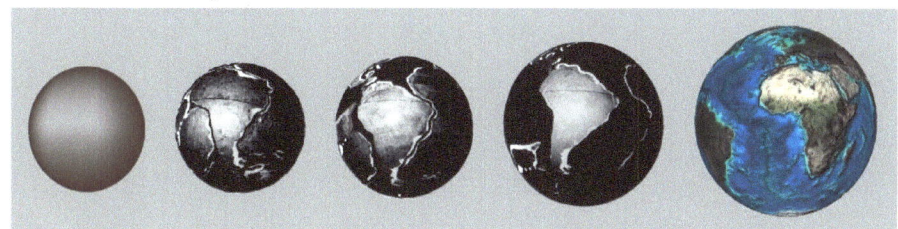

Figure 30. Decompression of Earth (WEDD) from Hadean to present.

In Figure 30 from left to right, same scale: 1) At Hadean, 64% of present Earth diameter, fully covered with continental-rock crust; 2), 3), & 4) Formation of primary and secondary decompression cracks that progressively fractured the continental-rock crust and open ocean basins. Timescale are not precisely established; 5) Holocene Earth.

I have unified the very-flawed Earth expansion theory and plate tectonics theory (which is dependent upon physically impossible mantle convection) into a whole new geoscience paradigm called Whole-Earth Decompression Dynamics (WEDD) [61] which identifies primary decompression cracks as underlain by heat sources, e.g., mid-ocean ridges, and secondary decompression cracks without heat sources, e.g., oceanic trenches. Primary decompression cracks, with their underlying heat sources, extrude basalt-rock, whereas secondary decompression cracks lack heat sources and become ultimate repositories for extruded basalt-rock. Basalt-rock, extruded from mid-ocean ridges, traverses the ocean floor by gravitational creep. Ultimately, in a process of "subduction" that lacks any mantle convection, seafloor basalt, with its carbonate sediment, fills in secondary decompression cracks, often located adjacent to continents. Seismically imaged "down-plunging slabs", I submit, are in-filled secondary decompression cracks.

Whole-Earth Decompression Dynamics extends plate tectonic concepts as it is responsible for Earth's well-documented features [59,61,64,121]. Fold-mountain formation does not necessarily require plate collisions [64], but rather is a consequence of crustal adjustments to curvature changes caused by volume increases. Partially in-filled secondary decompression cracks uniquely explain oceanic troughs, inexplicable by plate tectonics. And, compression heating at the base of the rigid crust is a direct consequence of mantle decompression [63]. Plate tectonic meanings and terminology are to a great extent preserved in my new paradigm. For example, transform plate boundaries are identical; divergent plate boundaries are similar, but with a different driving mechanism; convergent plate boundaries likewise are similar, but down-plunging plates neither create oceanic trenches, which are secondary decompression cracks, nor are they recycled through the mantle by conveyer-like mantle convection, and; Wadati-Benioff zones are quite similar, with the possible exception of why mantle melting occurs that is responsible for sometimes associated volcanic eruptions. Many of the plethora of observations, taken to support plate tectonics, support Whole-Earth Decompression Dynamics as well.

One aspect of science that the model-making, NASA-funded investigators never seem to realize is this: The objects in the Solar System came about through a series of causally related processes. As the nature of those processes becomes revealed through their consequences on meteoritic matter, logical and causal relationships lead to understanding; the pieces of the jigsaw puzzle fit smoothly together.

Implications from ordinary chondrites have adversely impacted geoscience without the origin of ordinary chondrites ever having been known [108]. For decades, the abundances of major elements in chondrites have been expressed in the literature as ratios, usually relative to silicon (E/Si) and occasionally relative to magnesium (E/Mg). By expressing Fe-Mg-Si elemental abundances as atom (molar) ratios relative to iron (E/Fe), as shown in Figure 31, I discovered a fundamental relationship bearing on the origin of ordinary chondrite matter [123].

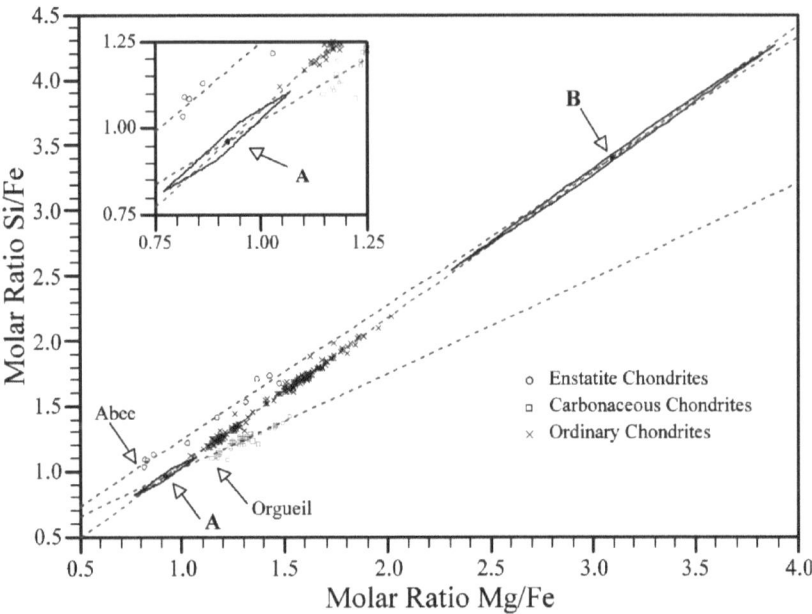

Figure 31. Atom (molar) ratios of Mg/Fe vs. Si/Fe from analytical data on 10 enstatite chondrites, 39 carbonaceous chondrites, and 157 ordinary chondrites.

In Figure 31 the well-defined lines are evident only when normalized to Fe, not to Si or Mg. The ordinary chondrite points scatter about a line that intersects the other two lines. Points on the ordinary chondrite line can be represented by mixtures of the two intersecting compositions, point A: *primitive*, and point B: *planetary*. The 95% confidence intervals are shown by solid lines only near points of intersection. For more detail, see [123].

The relationship I discovered admits the possibility of ordinary chondrites having been derived from mixtures of two components, representative of the other two types of matter, mixtures of a relatively undifferentiated carbonaceous-chondrite-like *primitive* component (point A) and a partially differentiated enstatite-chondrite-like *planetary* component (point B). ["Differentiated" here means that the element composition has been changed somewhat from primordial "solar" composition]. The planetary component, I posit, was the partially differentiated matter stripped from Mercury's protoplanet by the T-Tauri super-intense solar winds.

In a nutshell: The planets formed as gas giants, literally raining out from within giant gaseous protoplanets at high pressures. The super-intense T-Tauri outbursts associated with the ignition of the Sun stripped the gases from the inner planets and stripped away some of the matter from not-yet completed Mercury. That stripped away matter fused with low-pressure, low-temperature condensate, similar to the Orgueil carbonaceous chondrite, to form the parent matter of ordinary chondrites and most asteroids in the region of space between Mars and Jupiter. Viewed in that context, ordinary chondrite matter may be considered as a veneer that fell onto the outer portion of Earth, and to a greater degree, onto Mars.

Chapter 6. Science NASA Bungled

NASA's highly publicized Stardust Mission was designed to intercept a comet, acquire samples, and then return them safely to Earth for laboratory investigation, the expectation being that these samples would consist of "ancient pre-solar interstellar grains and nebula condensates that were incorporated into comets at the birth of the Solar System…." The target chosen was Wild 2, a comet discovered in 1978. This comet was thought to have spent most of its life at a greater distance from the Sun, but whose present orbit, shown in Figure 32, and orbital period of about six years was believed to have resulted from a 1974 gravitational interaction with Jupiter.

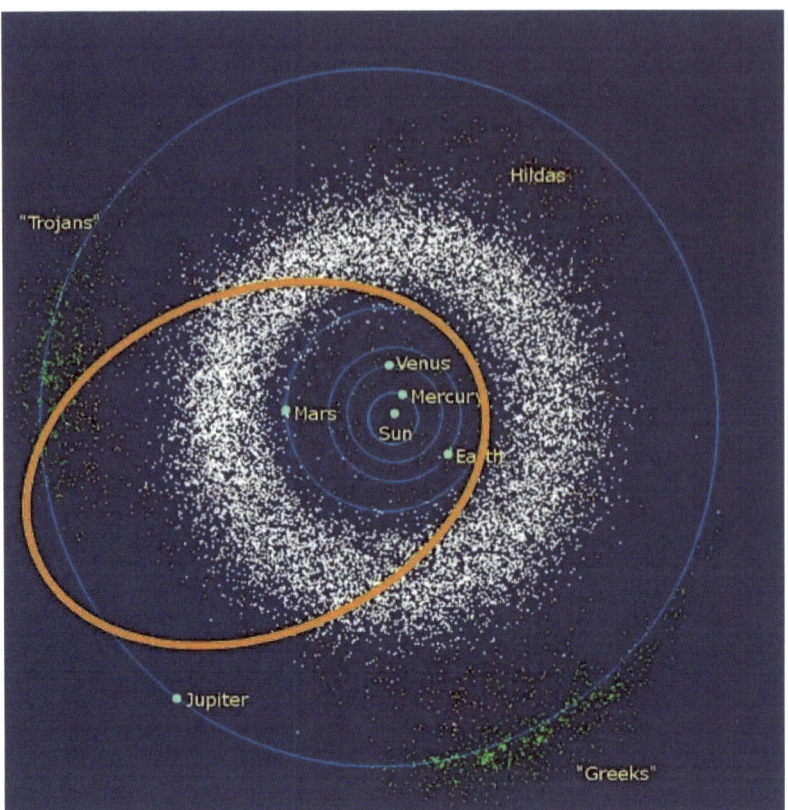

Figure 32. The orbit of Comet Wild 2 in April 2000 (orange) is shown for comparison to the orbits of Jupiter and the terrestrial planets.

In Figure 32 asteroids of the main asteroid belt are shown in white. Note that in recent times, at least since its discovery in 1978, Comet Wild 2 has repeatedly traversed the asteroid belt which lies between the orbits of Mars and Jupiter.

NASA-funded scientists have had experience trapping interstellar particles on silica gel in previous rocket and space shuttle flights. A similar technique was used to trap released particles from the comet's coma. But what NASA-funded scientists apparently failed to realize is this: In making repeated orbital revolutions almost entirely within the asteroid belt, the comet itself acted like a massive silica gel particle-trap, sampling and collecting particles it encountered in its sojourn through the asteroid belt. For the Stardust Mission to have achieved the intended results, it should have sampled a comet that had never entered the region of the asteroid belt where the comet would invariably have scavenged and trapped whatever debris was present. That was NASA's design flaw. Where was NASA's Chief Scientist? Out to lunch?

The Stardust spacecraft was launched February 7, 1999 and flew by comet Wild 2 on January 2, 2004. The sample-return canister landed back on Earth on January 15, 2006. It is not surprising, therefore, that crystalline particles similar to the minerals of ordinary chondrites and the re-melted/re-evaporated carbonaceous chondrites, for example, the mineral olivine were trapped.

The Stardust results might have been anticipated from a discovery made 1970. The Orgueil carbonaceous chondrite, which fell in France in 1864, has a state of oxidation and mineral components with characteristics similar to those which might be expected as a condensate from solar matter at low pressures, and some have supposed that it might have previously been the core of a comet. The major silicate is not a well-defined crystalline phase like olivine, but rather a poorly-characterized layer-lattice, clay-like, hydrous material. But small amounts of sharp, angular shards of crystalline olivine and pyroxene have in fact been found in that meteorite [124]. These shards, which show no alteration, appear to be a foreign, admixed component, perhaps originating in the asteroid belt.

Without realizing it beforehand, NASA spent 212 million dollars validating my view of the nature of the matter within the asteroid belt [125], but, to my knowledge, NASA never acknowledged that fact.

The September 30, 2011 issue of Science magazine was devoted to reports by the Project MESSENGER science team members of that NASA mission to orbit Mercury. For twenty years and continuing in the present, NASA-funded scientists, including Sean C. Solomon, the Principal Investigator for NASA's MESSENGER mission, have systematically and totally failed to mention or cite my scientific publications that bear on Mercury's origin, internal heat source, mode of magnetic field generation, and low-FeO content of Mercury's surface material that resembles the silicate-rock portion of an enstatite chondrite [53-58,62,75,90,123,126,127]. The September 30, 2011 Science reports were no exception; there was no reference to any of my relevant scientific publications.

The MESSENGER team members act as if they live in a gated community where they play at science, their science and their friend's science is permitted, but not my science. Playing at science for a NASA mission that cost taxpayers 446 million dollars! Hello! Science is not the exclusive domain of the NASA-funded in-crowd. Science is about telling the truth, the full truth. Those who practice deception in science, I have observed, generally fail to understand what science is all about and are often unable to explain scientific observations. Here is a case where MESSENGER scientists were unable to explain one of the most important observations made. Important images of Mercury's surface show features that had never before been observed on any other rocky planet or rocky moon. MESSENGER scientists were unable to explain those observations, but I could and did [128].

Many of the images from the MESSENGER spacecraft reveal "... an unusual landform on Mercury, characterized by irregular shaped, shallow, rimless depressions, commonly in clusters and in association with high-reflectance material ... and suggest that it indicates recent volatile-related activity" (Figures 33 and 34) [129]. But the authors were unable to describe a scientific basis for the source of those volatiles, stating: "...Mercury's interior contains higher abundances of volatile elements than are predicted by several planetary formation models for the innermost planet." Furthermore, the authors were unable to suggest identification of the "high-reflectance material".

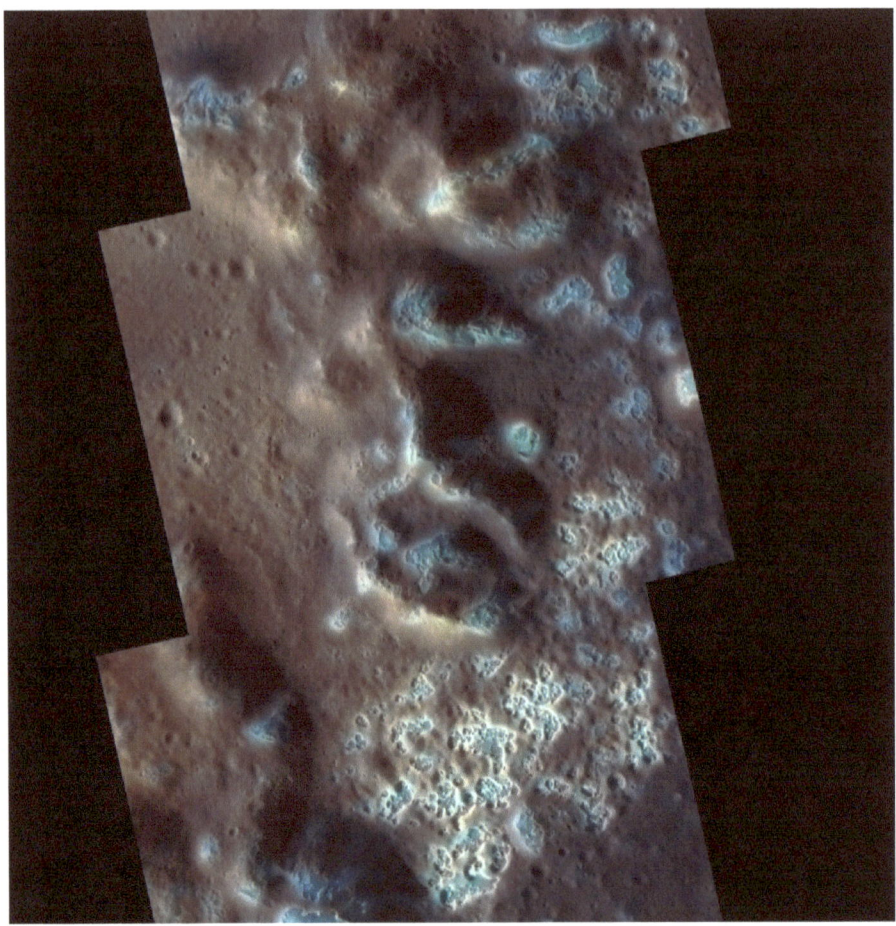

Figure 33. Colorized NASA MESSENGER image, taken with the Narrow Angle Camera, shows an area of hollows (blue) on the floor of Raditladi basin on Mercury.

In Figure 33 surface hollows were first discovered on Mercury during MESSENGER's orbital mission and have not been seen on the Moon or on any other rocky planetary bodies. These bright, shallow depressions appear to have been formed by disgorged volatile material(s) from within the planet.

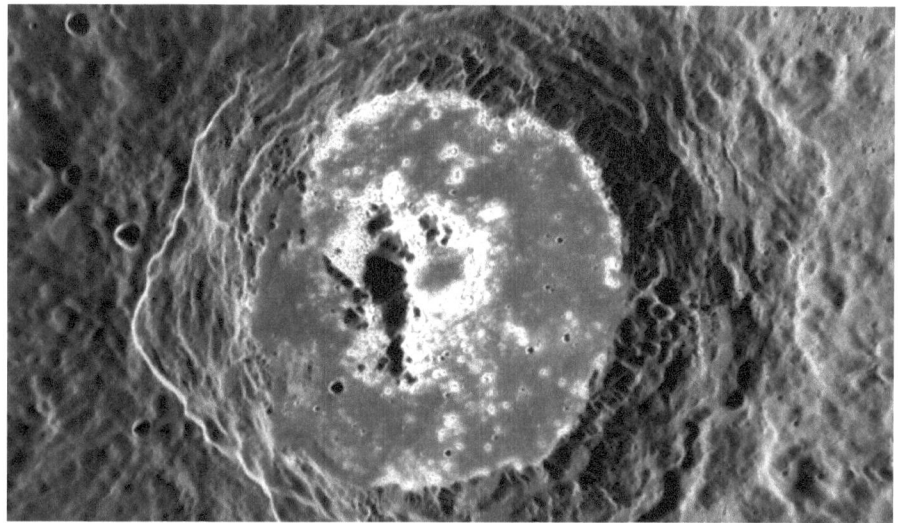

Figure 34. NASA MESSENGER image of a complex crater exhibits many hollows along its floor and central peak complex. The hollows have a very high albedo, which makes this crater stand out prominently.

I immediately understood those observations and described their basis in a scientific communication entitled "Hydrogen geysers: Explanation for observed evidence of geologically recent volatile-related activity on Mercury's surface" [128]. I provided the background on the my concept of Mercury formation by raining out from within a giant gaseous protoplanet at hydrogen pressures in excess of one atmosphere and the subsequent stripping away of primordial gases.

I showed that primordial condensation from an atmosphere of solar composition, at pressures above about one atmosphere, leads to iron metal condensing as a liquid. It is well known in metallurgy that molten iron is capable of dissolving copious amounts of hydrogen. My calculations indicate that one or more Mercury-size volumes of hydrogen (at STP) could potentially dissolve in the iron that rained-out to form Mercury's core, which is later released as the core solidifies.

The release of hydrogen escaping at the surface, I posited, is responsible for the formation of said "unusual landform on Mercury", sometimes referred to as pits, and for the formation of the associated "high-reflectance material", bright spots, which I suggested is iron metal reduced from an iron compound, probably iron sulfide, by the escaping hydrogen. The release of dissolved hydrogen during Mercury's core solidification is certainly sufficient in amount to account for the "unusual landform" on Mercury's surface and is possibly involved in the exhalation of iron sulfide, which is abundant on the planet's surface, and some of which may have been reduced by the hydrogen to iron metal thus accounting for the associated "high-reflectance material", bright spots.

Here is a test: Proving that the "high-reflectance material" is indeed metallic iron will provide strong evidence that the exhausted gas is hydrogen and evidence of the basis of planetary formation at pressures at or above about one atmosphere as I have described [62].

I have suggested that only three processes, operant during the formation of the Solar System, are primarily responsible for the diversity of matter in the Solar System and are directly responsible for planetary internal compositions and structures [62]. These are:

- (1) High-pressure, high-temperature condensation from primordial matter associated with planetary-formation by raining out from the interiors of giant-gaseous protoplanets;

- (2) Low-pressure, low-temperature condensation from primordial matter in the remote reaches of the Solar System or in the interstellar medium; and,

- (3) Stripping of the primordial volatile components from the inner portion of the Solar System by T-Tauri outbursts, presumably during the thermonuclear ignition of the Sun. This gives reason to understand the commonality of matter within planetary interiors and the georeactor-like nuclear fission generation of their magnetic fields.

Verifying that Mercury's "high-reflectance material" is indeed metallic iron will not only provide strong evidence for Mercury's hydrogen geysers, but more generally will provide evidence that planetary interiors "rained out" by condensing at high pressures within giant gaseous protoplanets [62].

In 2013 I published a General Article entitled "New Indivisible Planetary Science Paradigm" that that puts together all of my planetary discoveries and thus greatly supersedes NASA's very-flawed storyline [65].

Chapter 7. Deception at NASA Headquarters

Deception extends to the top-tier of NASA management. In a letter dated October 31, 2011, I made a formal complaint to NASA's Administrator, Maj. Gen. Charles F. Bolden, Jr., USMC, rtd. (Figure 40) that began: "I bring to your attention institutional corruption within the NASA establishment that has persisted for at least two decades, and that has diminished NASA's scientific credibility, viz., the publication of scientific misinformation and the concomitant deception of the scientific community and the general public, most recently regarding the interpretation of MESSENGER results published in the September 30, 2011 issue of Science."

Figure 35. NASA Administrator Maj. Gen. Charles F. Bolden, Jr., USMC, Retired.

In that letter and in its attached exhibits I provided substantial background for my allegations and noted: "The citation record can be readily verified through the Science Citation Index."

In that letter, I made the following three requests:

- (1) I herewith request that you initiate a full and open investigation into my allegations of science corruption at NASA and among NASA-supported individuals and institutions.

- (2) I herewith request that you insist that NASA scientists investigate without delay the bright, highly reflective matter observed associated with said [Mercury's] surface features, promptly make the results public, and, if said matter is indeed iron metal, acknowledge my explanation.

- (3) I request that NASA conspicuously and publically acknowledge my contributions to U. S. Space Science that I have made and which have been systematically ignored by NASA.

In mid-December, 2011, more than one month later, I received a response from NASA's Chief Scientist, Waleed Abdalati who stated: "… I understand your concerns, they are focused directly on the editorial practices of independent publications." Hello! That is a misrepresentation! Scientists themselves decide which publications to cite or not to cite.

Abdalati went on to state: "…NASA will not be investigating the practices of its investigators in this matter".

So the question becomes this: Did NASA Chief Scientist Abdalati misrepresent to conceal long-standing gross malfeasance of his office or did Abdalati misrepresent at NASA Administrator Bolden's behest?

What should have been NASA's Chief Scientist's response? In my view, Abdalati should have: (1) Initiated an independent investigation that involved as well the Federal Bureau of Investigation, and; (2) Begun to initiate programs that would permit NASA to publicly acknowledge my relevant, published advances in the planetary sciences that have been systematically ignored by scientists involved in investigations for NASA.

On December 20, 2011, I again wrote to NASA Administrator Bolden and stated: "So that there can be no question, I restate in words that should leave no doubt as to their meaning: For more than a decade, I allege, a coterie of scientists involved in investigations for NASA … have knowingly and willingly deceived NASA officials, the scientific community, and the general public, by systematically failing to mention or cite my relevant, published advances in the planetary sciences including, but not limited to, the feasibility of natural nuclear reactors as heat sources and generators of magnetic fields in planets and in large moons…."

I received no further response throughout the tenure of NASA Administrator Bolden. Am I surprised? Not really. NASA officials did not take seriously the evidence of 'O-ring' problems until the Challenger disaster, with its loss of all personnel on board, was viewed in real time on televisions throughout the world. Now, on the subject of science deception, NASA officials appear content to continue to deceive the scientific community, the public, teachers and students throughout the world.

As I wrote to NASA Administrator Maj. Gen. Charles F. Bolden, Jr., USMC, rtd.: "Science is all about truth and integrity, subjects the importance of which you were trained to appreciate [at the United States Naval Academy]: 'Midshipmen are persons of integrity: They stand for what is right. They tell the truth and ensure that the full truth be known. They do not lie.' Respectfully, in the matter at hand, your own integrity and leadership are on the line." I might have added, so is the integrity of the United States of America.

In an *Eos* interview [42], published April 18, 2018, former NASA Administrator Bolden was asked what advice he had for President Trump's newly appointed NASA Administrator, Jim Bridenstine, to which Bolden responded: "… Do your best to be nonpartisan enough to just put everything that you thought before on the back burner and become dedicated to the mission of the agency. Don't try to transform it, because it's been around a long time. It may need some tweaks, but it does not need to be remade."

Wrong! NASA needs to be eviscerated. NASA's deceitful science is not science at all; it is a storyline, a very-flawed storyline that misleads the public and minimizes America's standing on the world science-stage. NASA has no business "interact[ing] with more than 120 countries around the world" and NASA has no right to be party to an undisclosed international agreement, aiding and abetting the covert, systematic poisoning of the air we breathe, and the creatures of this planet breathe. NASA, which has long been immersed in its own politically-based obsessions, should have no involvement with weather and/or climate activities as those are and have been at the heart of the biggest international science-based scam ever perpetrated.

Appendix: Corruption of Science in America

Corruption of Science in America
J. Marvin Herndon, Ph.D.
Reprinted with Permission from
Dot Connector Magazine [1]

"That man will be very fortunate who, led by some unusual inner light, shall be able to turn from the dark and confused labyrinths within which he might have gone forever wandering with the crowd and becoming ever more entangled. Therefore, in the matter of philosophy, I consider it not very sound to judge a man's opinion by his followers." - Galileo Galilei, The Assayer (1623)

Truth is the pillar of civilization. The word 'truth' occurs 224 times in the King James Version of the Holy Bible; witnesses testifying in American courts and before the United States Congress must swear to tell the truth; and, laws and civil codes require truth in advertising and in business practices, to list just a few examples.

The purpose of science is to discover the true nature of Earth and Universe and to convey that knowledge truthfully to people everywhere. Science gives birth to technology that makes our lives easier and better. Science improves our health and enables us to see our world in ways never before envisioned. It uplifts spirits and engenders optimism. And, science provides a truth standard, securely anchored in the properties of matter, a means to expose and debunk the charlatans and science-barbarians who would lie, cheat, steal, and tyrannize under the guise of science.

Prior to World War II there was little government financial support for science. Nevertheless, the 20th century opened and seemed to offer the promise of an unparalleled age of enlightenment and reason.

While supporting himself as a patent clerk, Albert Einstein explained Brownian motion, the photoelectric effect, and special relativity. Niels Bohr, supported by grants from the Carlsberg Brewery, made fundamental discoveries about atomic structure and served as a focal point and driving force for the collaborative effort that yielded quantum mechanics, the field of science underpinning the solid state electronics technology that makes possible modern communications and computers. For a time, the meanings of new observations were actively debated. Fertile imaginations put forth ideas that challenged prevailing views. New ideas and new understandings began to emerge, sometimes precise; sometimes flawed, but tending toward truth and inspiring more new ideas and inspiring yet further debate. Individual imagination and creativity, driven by the quest for a true understanding of the nature of Earth and Universe, produced a sense of enthusiasm and excitement; new insights and discoveries enlightened the general public and kindled the imaginations of the young. An air of optimism prevailed.

Although money for science at the time was in short supply, scientists maintained a kind of self-discipline. A graduate student working on a Ph.D. degree was expected to make a new discovery to earn that degree, even if it meant starting over after years of work because someone else made the discovery first. Self-discipline was also part of the scientific publication system. Prior to World War II, when a scientist wanted to publish a paper, the scientist would send it to the editor of a scholarly journal for publication and generally it would be published. A new, unpublished scientist was required to obtain the endorsement of a published scientist before submitting a manuscript. The concept of 'peer review' had not yet been born.

But in the final decades of the 20th century, circumstances began to change. On one hand, outwardly, it seemed we were poised for yet another renaissance, with ready access to powerful new computers, satellite imaging, network data systems, and global communications. But, on the other hand, out of sight and unknown to nearly everyone, something had gone seriously wrong. Beneath the surface lay the foundations of a system which had been corrupted and had evolved to support a 'politically correct' consensus view of Earth and Universe, while tending to discourage, ignore, stifle and suppress advances and challenges by individuals.

Before World War II, there was very little government funding of science, but that changed because of war-time necessities. In 1951, the U.S. National Science Foundation (NSF) was established to provide support for post-World War II civilian scientific research. The process for administrating the government's science funding, invented in the early 1950s by NSF, has been adopted, essentially unchanged, by virtually all subsequent U.S. Government science funding agencies, such as the National Aeronautics and Space Administration (NASA) and the U.S. Department of Energy (DOE).

The problem, I discovered, is that the science funding process that the NSF invented and passed on to other U.S. Government agencies is seriously and fundamentally flawed. As a consequence, for more than half a century, the NSF has been doing what no foreign power or terrorist organization can do: slowly, imperceptibly undermining American scientific capability, driving America toward third-world status in science and in education, corrupting individuals and institutions, rewarding the deceitful and the institutions that they serve, stifling creative science, and infecting the whole scientific community with flawed anti-science practices based upon an unrealistic vision of human behavior. These are the principal flaws:

NSF Flaw #1: Proposals for scientific funding are generally reviewed by anonymous 'peer reviewers'. NSF invented the concept of 'peer review', where in a scientist's competitors would review and evaluate his/her/their proposal for funding, and the re-viewers' identities would be concealed. The idea of using anonymous 'peer reviewers' must have seemed like an administrative stroke of genius because the process was adopted by virtually all government science funding agencies that followed and almost universally by editors of scientific journals. But no one seems to have considered the lessons of history with respect to secrecy. Secrecy is certainly necessary in matters of national security and defense. But in civilian science, does secrecy and the concomitant freedom from accountability really encourage truthfulness? If secrecy did in fact lead to greater truthfulness, secrecy would be put to great advantage in the courts. Courts have in fact employed secrecy – during the infamous Spanish Inquisition and in virtually every totalitarian dictatorship – and the result is always the same: Unscrupulous individuals falsely denounce others and corruption abounds.

The application of anonymity and freedom from accountability in the 'peer review' system gives unfair advantage to those who would unjustly berate competitor's proposal for obtaining funding for research and for publishing research results. Anonymous 'peer review' has become the major science suppression method of the science-barbarians. Moreover, the perception – real or imagined – that some individuals would do just that has had a chilling effect, forcing scientists to become defensive, adopting only the 'politically correct' consensus approved viewpoint and refraining from discussing anything that might be considered a challenge to others' work or to the funding agency's programs. And that is not what science is about at all. Not surprisingly, there exists today a widespread perception that to challenge scientific results supported by a U.S. Government agency will lead to loss of one's own support.

NSF Flaw #2: NSF invented the concept of scientists proposing specific projects for funding, which has led to the trivialization and bureaucratization of science. Why so? The problem is that it is absolutely impossible to say beforehand what one will discover that has never before been discovered, and to say what one will do to discover it. The consequence has been the proposing of trivial projects with often non-scientific end--results, such as the widespread practice of making models based upon assumptions, instead of making discoveries. Further, bureaucrat 'program managers' decide which projects are suitable for the programs that they design. Moreover, proposal 'evaluation' is often a guise for 'program managers' and 'peer reviewers' to engage in exclusionary and ethically questionable, anti-competitive practices. There is no incentive for scientists to make important discoveries or to challenge existing ideas; quite the contrary.

NSF Flaw #3: NSF began the now widespread practice of making grants to universities and other non-profit institutions, with scientists, usually faculty members, now being classed as 'principal investigators'. The consequence of that methodology is that there is no direct legal responsibility or liability for the scientists' conduct. All too often scientists misrepresent with impunity the state of scientific knowledge and engage in anti-competitive practices, including the blacklisting of other capable, experienced scientists. University and institution administrators, when made aware of such conduct, in my experience, do nothing to correct it, having neither the expertise nor, with tenure, the perception of authority or responsibility. The result is that American tax payers' money is wasted on a grand scale and the science produced is greatly inferior to what it might be.

NSF Flaw #4: NSF began the now widespread practice whereby the government pays the publication costs, 'page charges', for scientific articles in journals run by for-profit companies or by special interest science organizations. Because these publishers demand ownership of copy rights, taxpayers who want to obtain an electronic copy must pay, typically US$40, for an article whose underlying research and publication costs were already paid with taxpayer dollars. Moreover, commercial and protectionist practices often subvert the free exchange of information, which should be part of science, making the publication of contradictions and new advances extremely difficult.

Furthermore, publishers have little incentive or mechanism to insist upon truthful representations. For example, in ethical science, published contradictions should be cited, but with the extant system it is common practice to ignore contradictions that may call into question the validity of what is being published. The net result is that unethical scientists frequently deceive the general public and the scientific community, and waste tax payer-provided money on questionable endeavors.

I have described these four fundamental NSF instigated flaws that now pervade virtually all civilian U.S. Government-supported science funding, and have proposed practical ways to correct them [130], which I communicated to two NSF directors, who chose to ignore them. There seems to be a widespread perception of intrinsic 'infallibility' in the government-university complex, where in any action, regardless of the seriousness of its adverse consequences, is considered beyond reproach.

On December 16, 2004, an individual in the White House to whom I had complained about the inequity of 'peer review' sent me a copy of the U.S. Office of Management and Budget's Final Information Quality Bulletin for Peer Review: December 15, 2004. On December 26, 2004, I sent to the White House my critique of that Bulletin and my recommendations for systemic changes, which were neither appreciated nor implemented [131]. Six years later, the U.S. Government still conducts 'peer review' according to that Bulletin, which: (1) Embodies the tacit assumption that 'peer reviewers' will always be truthful, and fails to provide any instruction, direction, or requirement either to guard against fraudulent 'peer review' or to prosecute those suspected of making untruthful reviews; (2) Approves the application of anonymity and even appears to promote some alleged virtue of its use, "e.g., to encourage candor"; (3) Gives tacit approval to circumstances that allow conflicts of interest and prevents the avoidance of conflicts of interest; and, (4) Fails to recognize or to admit the debilitating consequences of the long-term application of the practices it approves.

One consequence of NSF's invention of anonymous 'peer review' is that publication of scientific papers is often delayed for years or prevented by so-called 'peer reviews' from competitors, whose primary aim is to debilitate or eliminate their competition. In the 1990s, the National Science Foundation funded the development at Los Alamos National Laboratory of an author self-posting archive, where physicists and mathematicians could post their preprints, without interference from their competitors, making them available worldwide almost instantly. That archive underwent various name changes, eventually becoming arXiv.org.

Since its inception, arXiv.org has become the preeminent means of scientific communication in the areas of science and mathematics it hosts. Rather than wade through the many hundreds of individual scientific journals, often having limited access without paying fees, scientists can receive by email a list of daily postings in specific areas of the scientific disciplines hosted by arXiv.org and can download scientific articles of interest without charge. The development of the author self-posting archive might have become the jewel in NSF's crown, one of its greatest achievements. Instead, NSF's mal-administration permitted it to become an instrument for science suppression, and for blacklisting and discrimination against competent, well-trained scientists worldwide.

On or about 2001, key personnel responsible for developing the author self-posting archive at Los Alamos National Laboratory left that organization to become employed by Cornell University. Presumably in a coordinated way, Cornell University, through a proposal to the National Science Foundation [NSF # 0132355, July 16, 2001], took over ownership of the author self-posting archive, now called arXiv.org, and presumably was given the requested US$958,798 to do that. That proposal contains the following statement made to justify Cornell University's proposed use of a 'refereeing mechanism': "The research archives become less useful once they are inundated for example by submissions from vociferous 'amateurs' promoting their own perpetual motion machines...."

The website archivefreedom.org displays case histories of some of the individuals who have been blacklisted by the arXiv.org administration and its 'secret moderators', and includes a statement by blacklisted scientist and Nobel Laureate Brian D. Josephson explaining the meaning of blacklisting as applied to arXiv.org [132]. Being blacklisted by arXiv.org means that either your attempts to post scientific papers are disallowed, or they are 'buried', i.e., posted in categories where scientists or mathematicians in the specific area will likely not see them, such as in General Physics or in General Mathematics. The principal consequence of arXiv.org blacklisting is to deceive U.S. Government science funding officials and individuals conducting scientific investigations and teaching science, keeping them in the dark about new ideas and discoveries. Beyond the financial and professional debilitation suffered by blacklisted scientists and mathematicians, there is also a human toll, as one blacklisted individual noted: "Blacklisted scientists are subject to derision, ignorance, insults, lies, false accusations, personal attacks against them, misrepresentations regarding their research, culture, faith, etc."

Hundreds of thousands of scientific papers have been posted on the author self-posting archive, arXiv.org, without any human intervention at all. Human intervention, but not 'peer review', occurs only when an individual is 'denounced', intentionally singled out for disparate treatment, through the application of unfair, arbitrary, and capricious standards. Being tagged for disparate human intervention may occur for a number of never specified reasons. Human intervention is perpetrated by arXiv.org administrators in conspiracy with a small group of arXiv.org 'insiders' who may or may not call themselves 'moderators' and who discriminate in secret and without any accountability. Moreover, there is no recourse: in my experience, Cornell University's librarian, provost and president absolve themselves from any oversight responsibility for the conduct of arXiv.org, referring complaints back to the arXiv.org administrators who are the subject of the complaint in the first place. Being 'denounced' for disparate treatment by secret 'insiders', without recourse, is something I might have expected from the now-defunct Soviet Union or from Ceausescu's Romania. But, here it is in America; bought and paid for by the National Science Foundation. As an American citizen, veteran, and taxpayer, I am justifiably appalled!

In my view, there is something fundamentally wrong with Cornell University receiving U.S. Government grants and contracts to conduct scientific research, and then deceiving the scientific community, via arXiv.org, by not posting or by hiding new advances or contradictions, especially in instances that potentially impact the investigations being performed at government expense at Cornell. Cornell University is a recipient of millions of dollars in U.S. Government grants and contracts, and is one of a pool of competitors for Federal grants and contracts. The National Science Foundation, I submit, made an institutionally stupid blunder in turning over to Cornell University a powerful tool (arXiv.org) that could be used against its competitors. In doing so, I allege, the U.S. National Science Foundation violated the very law that created NSF: "In exercising the authority and discharging the functions referred to in the foregoing subsections, it shall be an objective of the Foundation to strengthen research and education in the sciences and engineering, including independent research by individuals, throughout the United States, and to avoid undue concentration of such research and education." [42 United States Code 1862 (e)]

Instead of obeying that law, the U.S. National Science Foundation placed into the hands of one major, well-financed competitor a powerful tool (arXiv.org) which could not only be applied arbitrarily with capricious standards against its competitors, but through such actions would cast a shadow of fear at being 'denounced' in secret and there upon being blacklisted, further ensuring 'politically correct' consensus conformity and science suppression. So, what should be done?

In my view, the United States Congress should initiate an investigation into allegations of abuse and possible criminal activity in the acquisition and operation of arXiv.org at Cornell University, including the possibility of complicity and/or acquiescence by individuals at other universities and by other government entities, including the U.S. Department of Justice and the Attorney General of the State of New York. If evidence warrants, the United States Government, I believe, should consider initiating legal action to repossess arXiv.org and put it under aegis of a neutral, non-competitor organization, such as the National Archive or the Library of Congress, as should have been done initially.

The noted economist, George E. P. Box, said essentially this about models [2]: all models are wrong, but some are useful. Generally, models set out to model some observable or hypothetical event or process and achieve the result they seek to obtain by making result-oriented assumptions and tweaking variables; those models do not have to be correct and can generally be replaced with other models. To me, it is much more important to discover the true nature of Earth and Universe than to make such models.

Astronomers have made some truly remarkable observations. Astrophysicists attempt to understand the physical basis underlying those observations by making models based upon assumptions or upon other models based on other assumptions. In the 1920s, scientists discovered thermonuclear fusion, the joining of two very light atomic nuclei with great energy release. The process is called 'thermonuclear' because temperatures of about one million degrees centigrade are required to ignite the reaction. In the 1930s, scientists worked out the thermonuclear reactions thought to power the Sun and other stars. The million degree ignition temperature? It was assumed to be generated when dust and gas collapsed during their formation. But, as I realized later, there are serious impediments to attaining million degree temperatures in that manner.

A star is like a hydrogen bomb held together by gravity. The thermonuclear fusion reactions of all hydrogen bombs are ignited by small nuclear fission atomic bombs. In 1994, in a scientific paper published in the Proceedings of the Royal

Society of London, I suggested that stars, like hydrogen bombs, are ignited by nuclear fission, the splitting of uranium and heavier atomic nuclei [54]. The implications are profound: stars are not necessarily ignited during formation, as previously thought, but require a fissionable trigger. My concept of the thermonuclear ignition of stars by nuclear fission has been completely ignored by the model making astrophysicists. Ignoring work that challenges the 'politically correct' consensus-approved storyline is common practice, thanks to the fear of retribution by secret 'peer reviewers' or to the fear of being 'denounced' and blacklisted.

In 2006, I submitted a short manuscript on the thermonuclear ignition of dark galaxies to Astrophysical Journal Letters. I signed the required copyright transfer form, and the manuscript went out for secret 'peer review', but it was rejected without any substantive scientific criticism. So I submitted two other brief, but important, manuscripts. The fact that I was never asked to sign the copyright transfer forms for those other two papers prior to review, as required, was clear indication that they were not going to be accorded the fair and impartial consideration that is supposed to be the usual policy of the American Astronomical Society, the journal's sponsor. Not surprisingly, those manuscripts were rejected without any scientifically valid justification. I complained to the officers of the American Astronomical Society, who never responded, even though the bylaws of the American Astronomical Society (AAS) clearly state: "As a professional society, the AAS must provide an environment that encourages the free expression and exchange of scientific ideas." In rejecting those manuscripts, the American Astronomical Society hid from its members, from the scientific community, and from U.S. Government science funding officials, fundamentally new insights about the Universe, including why galaxies have the characteristic appearances they are observed to have [133].

Not long after the Astrophysical Journal Letters incident, I found myself blacklisted by arXiv.org. Before, I was not only permitted to post, but also to endorse others in the following categories: Astrophysics, Educational Physics, General Physics, Geophysics, History of Physics, and Space Physics. Now, for no legitimate reason, I am blacklisted, stripped of the ability to endorse others, and suffer having my scientific papers 'buried' in General Physics where it is unlikely they will be noticed; that is, if they are allowed to post at all. Even my scientific papers that call into question U.S. Government funded investigations at Cornell University are either 'buried' or forbidden to post in this author self-posting archive, where hundreds of thousands of papers post automatically without human intervention.

A half-century of the use of secret 'peer reviews' by competitors, at the National Science Foundation and at the other agencies which followed, such as NASA, has produced a 'never criticize the science' mentality among grant-recipients. But science is all about finding out what is wrong with present thinking and correcting it. American science education has been stunted by that mentality. Educational organizations which receive grants from NSF or NASA almost never teach students or teachers about work that challenges the 'politically correct' consensus approved storyline. The same goes for 'science news' organizations that rarely report the results of investigations that call into question the 'politically correct' story line. Institutionalized science-corruption is wide spread and pervasive in America, and the fallout is international; the 'Climategate' debacle is just one example.

At one time, scientists thought that planets do not produce energy, except small amounts from radioactive decay; planets just receive energy from the Sun and then radiate it back into space. Beginning in the late 1960s, astronomers observed that Jupiter, Saturn and Neptune radiate into space nearly twice the energy they receive from the Sun. For twenty years the source of that internal energy was a mystery to NASA-funded scientists, who wrongly thought they had considered and eliminated all possibilities. In 1991, I submitted a scientific paper to the German Naturwissenschaften demonstrating the feasibility of that energy being produced by natural nuclear fission reactors at the planets' centers. I used the same approach that Paul K. Kuroda had used in 1956 to predict the occurrence of natural nuclear reactors in ancient uranium mines, the fossil remains of which were discovered in 1972 at Oklo, in the Republic of Gabon.

When that paper was accepted for publication [52], I submitted a research proposal to NASA's Planetary Geophysics Program. Paul K. Kuroda accepted my invitation to join in as a co-investigator. Kuroda, however, insisted that his efforts be pro bono as he 'did not need the money'.

The Universities Space Research Association, an association of major institutional recipients of NASA funding, operates the Lunar and Planetary Institute, which operated the Lunar and Planetary Geoscience Review Panel (LPGRP) at the time I submitted the proposal. The LPGRP served NASA by soliciting secret 'peer reviews' of submitted proposals, then evaluating the proposals in secret session, based upon those 'peer reviews', and ranking them so as to make it easy for a NASA official to decide which to fund. The LPGRP, composed of a group of principal investigators of NASA grants, funded either through NASA's Planetary Geophysics Program or Planetary Geology Program, conducted the secret ranking of all proposals submitted to one or the other of those same two NASA programs. In other words, my proposal was competing for the same limited pool of funds as proposals from the very institutions whose personnel served on the LPGRP. At the time, the chairman of the LPGRP was associated with NASA's Jet Propulsion Laboratory, which is operated by the California Institute of Technology (Caltech), and which consumed more than 40% of the budget of the Planetary Geophysics Program.

Needless to say, my proposal was not funded. Normally, the LPGRP's ranking of proposals is kept secret, but through extraordinary efforts I learned from the U.S. Congress' General Accounting Office (called the Government Accountability Office since 2004) that on technical merit the LPGRP ranked my proposal lowest of the 120 proposals submitted to NASA's Planetary Geophysics Program. One might seriously question the integrity of that ranking, as I later independently performed all that I had proposed and much more, including demonstrating the feasibility of a nuclear fission reactor at the center of Earth, called the georeactor, as the energy source and production mechanism for the Earth's magnetic field [53,54,56,57,90]. I also extended the concept to other planets and large moons [58]. The concept of planetary nuclear fission reactors has received quite thorough vetting in the international scientific community. So, what was NASA's response?

In the twenty years that have passed since the proposal debacle, NASA supported scientists, to my knowledge, have never mentioned natural nuclear fission reactors or cited my publications. But they have discussed numerous observations where they should have, instances of 'mysterious' internal heat production and magnetic field generation, such as: (1) Internal heat generation in Jupiter, Saturn and Neptune; (2) Our Moon having a soft or molten core; (3) Tiny planet Mercury having a magnetic field; (4) Mars displaying evidence of an ancient magnetic field; (5) Our Moon displaying evidence of an ancient magnetic field; (6) Jupiter's moon Ganymede having an internally generated magnetic field; (7) Saturn's moon Enceladus showing evidence of internal heating; and (8) Evidence of internal heat generation in Pluto's moon Charon. I receive numerous emails from people throughout the world who read NASA news reports and wonder why my work is not mentioned, when it would seem to provide plausible explanations.

In a manner no different from astrophysics, the American geophysical community consistently ignores my scientific challenges to the 1940 vintage thoughts that form the basis of their assumption-based models. Science is not about telling one 'politically correct' story and ignoring everything else. Instead, science is about finding out what is wrong with existing ideas and correcting them. American geophysicists have wasted untold multi-millions of taxpayer provided dollars on totally worthless endeavors, instead of progressing in fruitful directions. I publish important, well-founded contradictions to current scientific thinking in world-class journals. It is the responsibility of an ethical scientific community to attempt to confirm or to refute the concepts presented. In any case, those contradictions should be cited [134].

In 1936, Inge Lehmann discovered the inner core, an object at the center of Earth almost as large as the Moon and about three times as massive, that, since about 1940, was thought to be iron in the process of freezing. In 1979, I published an entirely different idea of the inner core's composition. The scientific paper was communicated by Nobel Laureate Harold C. Urey to the Proceedings of the Royal Society of London [85] and I received a complimentary letter from Inge Lehmann. But instead of debate, discussion, and experimental and/or theoretical verification/ refutation, I received silence from the geophysics community, not only on that discovery, but on a host of discoveries that followed as a consequence [135]. Real scientists welcome new ideas and advances as they open the door to more new ideas and further advances. Science-barbarians, on the other hand, ignore what they do not like, and by ignoring, deceive the scientific community, the general public, and the U.S. Government, which typically funds their questionable endeavors.

In 1838, in an address before the Young Men's Lyceum of Springfield, Illinois, Abraham Lincoln stated: "At what point, then, is the approach of danger to be expected? I answer if it ever reaches us, it must spring up amongst us. It cannot come from abroad. If destruction be our lot, we must ourselves be its author and finisher." Later, U.S. President Abraham Lincoln unknowingly helped to sow the seeds for America's self-destruction when in 1863 he signed into law the Act of Incorporation of the National Academy of Sciences, which states in part: "The National Academy of Sciences shall… whenever called upon by any department of the Government, investigate, examine, experiment, and report upon any subject of science or art."

Has the National Academy of Sciences ever advised the U.S. Government of the flaws in the operating procedures of science-funding agencies, such as I have disclosed [130,131], which are corrupting and trivializing American science? Has it ever revealed the existence of organized science-suppression under the guise of secret 'peer review' among the so-called professional societies, including within the National Academy of Sciences, the documentation of which I have provided to the president of NAS, and the consequences of which will cost American taxpayers countless millions of wasted tax dollars? I doubt it. Despite ever-increasing budgets, American science and education continues to decline toward third-world status as it has for decades. In personal, medical, legal, and business matters, it is common practice to hire an advisor. We all do that. If the advice proffered proves to be faulty, we fire the advisor and hire another. In my opinion, the United States Congress should fire the U.S. National Academy of Sciences and find other sources of scientific and educational advice.

Suppressing and ignoring advances in science can have serious, real-world consequences. The Earth is constantly bombarded by the solar wind, a fully ionized and electrically conducting plasma, heated to about 1,000,000 °C. Fortunately, Earth's self--generated magnetic field deflects the brunt of the solar wind safely around and past our planet, protecting humanity from the Sun's relentless onslaught.

But reversal or demise of the geomagnetic field will doubtlessly be catastrophic, a calamity of unparalleled magnitude for our technologically-dependent civilization.

When the geomagnetic field collapses, vast segments of the population will be without electricity. Electrical power grids will act like uncontrolled generators as the charged -particle flux of the rampaging solar wind sweeps past, inducing into their lines suicidal bursts of electrical current that short-circuit and destroy essential elements of the power grid. Powerful, equipment-wrecking electrical currents will likewise be induced in gas and oil pipelines, causing explosions and fires. Electrical charges will build up on surfaces everywhere and reach staggeringly high potentials at edges and sharp points, posing risks of electrocution and igniting fires. Satellites will no longer function, their electronics fried by the plasma onslaught; there will be widespread failure of both communication and navigation systems. And, even more seriously, the long-term, unknown, but certainly adverse, impact on health will be severe.

Until recently, reversals of the geomagnetic field or its complete demise were thought to be events in the far distant future and to occur over a long period of time. But that may have changed dramatically.

Notice that as you heat a pot of water on the stove top, before it starts to boil, the water begins to circulate from bottom to top and from top to bottom. This is called convection and it can be better observed by adding a few tea leaves, celery seeds, or the like, which are carried along by the circulation of water. It occurs because heat at the bottom causes the water to expand a bit, becoming lighter, less dense, than the cooler water at the top. This process of convection is an unstable, top heavy arrangement which attempts to regain stability by fluid motions.

In 1939, Walter Elsasser proposed that the geomagnetic field is produced by convection motions in the Earth's fluid core that are twisted by the planet's rotation to form a dynamo. For seventy years, the geophysics community has assumed that convection 'must' exist in the core. Untold millions of dollars have been spent on modeling convection and its applications in the Earth's fluid core.

On January 27, 2009, I submitted a brief but important scientific communication to Physical Review Letters which demonstrated that convection is physically impossible in the Earth's fluid core because: (1) The core is too bottom-heavy due to compression by the weight above; (2) The core-bottom cannot remain hotter than the top, as required for convection, because the core is wrapped in an insulating blanket; and, (3) The 'Rayleigh Number' has been wrongly applied to justify core convection. I suggested instead that the geomagnetic field is produced by Elsasser's mechanism operating in the nuclear georeactor sub-shell. From bottom to top in the review process at Physical Review Letters and at the journal's sponsor, the American Physical Society, there were no scientifically valid, substantive criticisms, only pejorative remarks and misrepresentations, including those by one or more members of the National Academy of Sciences. Of course, the paper was rejected by Physical Review Letters and its pre-print was 'buried' by arXiv.org in General Physics [136], which effectively hid it from view of U.S. Government science funding officials, almost guaranteeing that fluid core modeling activities would continue wasting tax payer funds on fruitless, physically impossible endeavors. But there is a far, far more serious implication stemming from the unwarranted rejection and 'burial' of this manuscript.

Earth's fluid core comprises about 30% of the mass of the planet; the nuclear georeactor is only one ten-millionth as massive, meaning that disrupted convection in the georeactor could lead to very rapid changes, including rapid reversals of the geomagnetic field. Think of it this way: the direction and speed of a child's tiny, self-moving toy train can be changed much more rapidly with far less force than that of the longest and heaviest, fully loaded, full-size freight train. From ancient lava flows, scientists have recently confirmed evidence of episodes of rapid geomagnetic field change – six degrees per day during one reversal and another of one degree per week – were reported [137,138]. The relatively small mass of the georeactor is consistent with the possibility of a magnetic reversal occurring on a time scale as short as one month or several years. The recently observed more-rapid-than-usual movement of the North magnetic pole toward Siberia is thought by some to suggest that a reversal is imminent, although there is great uncertainty. Because of the global catastrophic significance, suppressing science related to the possibility of very rapid geomagnetic field changes, in my view, is tantamount to a betrayal of trust and an act of treason against humanity.

For the good of all, now is the time to rid science of the charlatans and the science -barbarians, and to create an environment where science can flourish in truth and where scientists can work freely without fear of retribution or denouncement for challenging extant ideas or for failing to adopt the 'politically correct' consensus-approved storyline. I have described four major, science-crippling flaws, instigated by the U.S. National Science Foundation a half-century ago, that are still in effect today at NSF, and at other U.S. Government science funding agencies, and have suggested practical ways to correct them [130,131]. Implementation should not be too difficult; it just requires courage and integrity.

References

1. Herndon, J.M. Corruption of Science in America. The Dot Connector 2011.
2. Box, G.E.P. Empirical model-building and response surfaces. Wiley: 1987.
3. Holmes, A. Radioactivity and earth movements. Trans. geol. Soc. Glasgow 1928-1929 1931, 18, 559-606.
4. Herndon, J.M. Geodynamic basis of heat transport in the earth. Curr. Sci. 2011, 101, 1440-1450.
5. Revelle, R.; Suess, H.E. Carbon dioxide exchange between atmosphere and ocean and the question of an increase of atmospheric CO2 during the past decades. Tellus 1957, 9, 18-27.
6. Herndon, J.M. Variables unaccounted for in global warming and climate change models. Curr. Sci. 2008, 97, 815-816.
7. Delingpole, J. Climategate: The final nail in the coffin of 'anthropogenic global warming'? The Telegraph 20 Nov 2009 2009.
8. Costella, J. The climategate emails. The Lavoisier Group: Australia, 2010.
9. Herndon, J.M. Evidence of variable earth-heat production, global non-anthropogenic climate change, and geoengineered global warming and polar melting. J. Geog. Environ. Earth Sci. Intn. 2017, 10, 16.

10. Herndon, J.M. An open letter to members of agu, egu, and ipcc alleging promotion of fake science at the expense of human and environmental health and comments on agu draft geoengineering position statement. New Concepts in Global Tectonics Journal 2017, 5, 413-416.

11. Sands, P. East west street: On the origins of" genocide" and" crimes against humanity". Knopf: 2016.

12. http://www.nuclearplanet.com/websites.pdf

13. Fleming, J.R. Fixing the sky: The checkered history of weather and climate control. Columbia University Press: New York, 2010.

14. Kampa, M.; Castanas, E. Human health effects of air pollution Environmental Pollution 2008, 151, 362-367.

15. Calderon-Garciduenas, L.; Franko-Lira, M.; Mora-Tiscareno, A.; Medina-Cortina, H.; Torres-Jardon, R.; et al. Early alzheimer's and parkinson's diese pathology in urban children: Friend verses foe response - it's time to face the evidence. BioMed Research International 2013, 32, 650-658.

16. Moulton, P.V.; Yang, W. Air pollution, oxidative stress, and alzheimer's disease. Journal of Environmental and Public Health 2012, 109, 1004-1011.

17. Beeson, W.L.; Abbey, D.E.; Knutsen, S.F. Long-term concentrations of ambient air pollutants and incident lung cancer in california adults: Results from the ahsmog study. Environ. Health Perspect. 1998, 106, 813-822.

18. Hong, Y.C.; Lee, J.T.; Kim, H.; Kwon, H.J. Air pollution: A new risk factor in ischemic stroke mortality. Stroke 2002, 33, 2165-2169.

19. Haberzetti, P.; Lee, J.; Duggineni, D.; McCracken, J.; Bolanowski, D.; O'Toole, T.E.; Bhatnagar, A.; Conklin, D., J. Exposure to ambient air fine particulate matter prevents vegf-induced mobilization of endothelial progenitor cells from bone matter. Environ. Health Perspect. 2012, 120, 848-856.

20. Potera, C. Toxicity beyond the lung: Connecting PM2.5, inflammation, and diabetes. Environ. Health Perspect. 2014, 122, A29.

21. Mehta, A.J.; Zanobetti, A.; Bind, M.-A., C.; Kloog, I.; Koutrakis, P.; Sparrow, D.; Vokonas, P.S.; Schwartz, J.D. Long-term exposure to ambient fine particulate matter and renal function in older men: The va normative aging study. Environ. Health Perspect. 2016, 124, 1353-1360.

22. Dai, L.; Zanobetti, A.; Koutrakis, P.; Schwartz, J.D. Associations of fine particulate matter species with mortality in the united states: A multicity time-series analysis. Environ. Health Perspect. 2014, 122, 837-842.

23. Dockery, D.W.; Pope, C.A.I.; Xu, X.P.; Spengler, J.D.; Ware, J.H.; et al. An association between air pollution and mortality in six u. S. Cities. N. Eng. J. Med. 1993, 329, 1753-1759.

24. Pope, C.A.I.; Ezzati, M.; Dockery, D.W. Fine-particulate air polution and life expectancy in the united states. N. Eng. J. Med. 2009, 360, 376-386.

25. Pires, A.; de Melo, E.N.; Mauad, T.; Saldiva, P.H.N.; Bueno, H.M.d.S. Pre- and postnatal exposure to ambient levels of urban particulate matter (PM2.5) affects mice spermatogenesis. Inhalation Toxicology: International Forum

for Respiratory Research: DOI: 10.3109/08958378.2011.563508 2011, 23.

26. Ebisu, K.; Bell, M.L. Airborne PM2.5 chemical components and low birth weight in the northeastern and mid-atlantic regions of the united states. Environ. Health Perspect. 2012, 120, 1746-1752.

27. Tetreault, L.-F.; Doucet, M.; Gamache, P.; Fournier, M.; Brand, A.; Kosatsky, T.; Smargiassi, A. Childhood exposure to ambient air pollutants and the onset of asthma: An administrative cohort study in quebec. Environ. Health Perspect. 2016, 124, 1276.

28. Bell, M.L.; Ebisu, K.; Leaderer, B.P.; Gent, J.F.; Lee, H.J.; Koutrakis, P.; Wang, Y.; Dominici, F.; Peng, R.D. Associations of PM2.5 constituents and sources with hospital admissions: Analysis of four counties in connecticut and massachusetts (USA). Environ. Health Perspect. 2014, 122, 138-144.

29. Herndon, J.M. Aluminum poisoning of humanity and earth's biota by clandestine geoengineering activity: Implications for india. Curr. Sci. 2015, 108, 2173-2177.

30. Herndon, J.M. Obtaining evidence of coal fly ash content in weather modification (geoengineering) through analyses of post-aerosol spraying rainwater and solid substances. Ind. J. Sci. Res. and Tech. 2016, 4, 30-36.

31. Herndon, J.M. Adverse agricultural consequences of weather modification. AGRIVITA Journal of agricultural science 2016, 38, 213-221.

32. Herndon, J.M.; Whiteside, M. Further evidence of coal fly ash utilization in tropospheric geoengineering: Implications on

human and environmental health. J. Geog. Environ. Earth Sci. Intn. 2017, 9, 1-8.

33. Whiteside, M.; Herndon, J.M. Coal fly ash aerosol: Risk factor for lung cancer. Journal of Advances in Medicine and Medical Research 2018, 25, 1-10.

34. Whiteside, M.; Herndon, J.M. Aerosolized coal fly ash: Risk factor for copd and respiratory disease. Journal of Advances in Medicine and Medical Research 2018, 26, 1-13.

35. Whiteside, M.; Herndon, J.M. Aerosolized coal fly ash: Risk factor for neurodegenerative disease. Journal of Advances in Medicine and Medical Research 2018, 25, 1-11.

36. Herndon, J.M.; Whiteside, M. Contamination of the biosphere with mercury: Another potential consequence of on-going climate manipulation using aerosolized coal fly ash J. Geog. Environ. Earth Sci. Intn. 2017, 13, 1-11.

37. http://www.nuclearplanet.com/public_rejection.pdf

38. http://www.nuclearplanet.com/USAF.pdf

39. http://www.nuclearplanet.com/Public_Deception_by_Scientists.html

40. http://www.nuclearplanet.com/explainretractions.pdf

41. http://www.nuclearplanet.com/nasacwk.pdf

42. Showstack, R. Former nasa administrator weighs in on new space agency head. Eos 2018, 99.

43. Peek, B.M. The planet jupiter. Faber and Faber: London, 1958.

44. Conrath, B.J.; Pearl, J.C.; Appleby, J.F.; Lindal, J.F.; Orton, G.S.; Bezard, B. In Uranus, Bergstralh, J.T.; Miner, E.D.; Mathews, M.S., Eds. University of Arizona Press: Tucson, 1991.

45. Hubbard, W.B. Interiors of the giant planets. In The new solar system, Chaikin, J.K.B.a.A., Ed. Sky Publishing Corp.: Cambridge, MA, 1990; pp 134-135.

46. Stevenson, J.D. The outer planets and their satellites. In The origin of the solar system, Dermott, S.F., Ed. Wiley: New York, 1978; pp 395-431.

47. Kuroda, P.K. On the nuclear physical stability of the uranium minerals. J. Chem. Phys. 1956, 25, 781-782.

48. Bodu, R.; Bouzigues, H.; Morin, N.; Pfiffelmann, J.P. Sur l'existence anomalies isotopiques rencontrees dan l'uranium gu gabon. C. r. Acad. Sci., Paris 1972, D275, 1731-1736.

49. Cowan, G.A. A natural fission reactor. Sci. Am. 1976, 235, 36-47.

50. Frejacques, C.; Blain, C.; Devilers, C.; Hagemann, R.; Ruffenbach, J.-C. In The oklo phenomenon, I.A.E.A.: Vienna, 1975; p 509.

51. Maurette, M. Fossil nuclear reactors. A. Rev. Nuc. Sci. 1976, 26, 319-350.

52. Herndon, J.M. Nuclear fission reactors as energy sources for the giant outer planets. Naturwissenschaften 1992, 79, 7-14.

53. Herndon, J.M. Feasibility of a nuclear fission reactor at the center of the earth as the energy source for the geomagnetic field. J. Geomag. Geoelectr. 1993, 45, 423-437.

54. Herndon, J.M. Planetary and protostellar nuclear fission: Implications for planetary change, stellar ignition and dark matter. Proc. R. Soc. Lond 1994, A455, 453-461.

55. Herndon, J.M. Sub-structure of the inner core of the earth. Proc. Nat. Acad. Sci. USA 1996, 93, 646-648.

56. Herndon, J.M. Nuclear georeactor origin of oceanic basalt 3He/4He, evidence, and implications. Proc. Nat. Acad. Sci. USA 2003, 100, 3047-3050.

57. Herndon, J.M. Nuclear georeactor generation of the earth's geomagnetic field. Curr. Sci. 2007, 93, 1485-1487.

58. Herndon, J.M. Nature of planetary matter and magnetic field generation in the solar system. Curr. Sci. 2009, 96, 1033-1039.

59. Herndon, J.M. Terracentric nuclear fission georeactor: Background, basis, feasibility, structure, evidence and geophysical implications. Curr. Sci. 2014, 106, 528-541.

60. Herndon, J.M. Examining the overlooked implications of natural nuclear reactors. Eos, Trans. Am. Geophys. U. 1998, 79, 451,456.

61. Herndon, J.M. Whole-earth decompression dynamics. Curr. Sci. 2005, 89, 1937-1941.

62. Herndon, J.M. Solar system processes underlying planetary formation, geodynamics, and the georeactor. Earth, Moon, and Planets 2006, 99, 53-99.

63. Herndon, J.M. Energy for geodynamics: Mantle decompression thermal tsunami. Curr. Sci. 2006, 90, 1605-1606.

64. Herndon, J.M. Origin of mountains and primary initiation of submarine canyons: The consequences of earth's early formation as a jupiter-like gas giant. Curr. Sci. 2012, 102, 1370-1372.

65. Herndon, J.M. New indivisible planetary science paradigm. Curr. Sci. 2013, 105, 450-460.

66. Herndon, J.M. New concept for the origin of fjords and submarine canyons: Consequence of whole-earth decompression dynamics. Journal of Geography, Environment and Earth Science International 2016, 7, 1-10.

67. Herndon, J.M. Fictitious supercontinent cycles. Journal of Geography, Environment and Earth Science International 2016, 7, 1-7.

68. Elkins-Tanton, L.T. Uranus, neptune, pluto, and the outer solar system. Facts on File, Inc.: New York, NY, 2011.

69. Elkins-Tanton, L.T. Jupiter and saturn. Infobase Publishing: New York, 2006; p 220.

70. Kivelson, M.G.; Khurana, K.K.; Walker, R.J.; Russell, C.T.; Linker, J.A.; Southwood, D.J.; Polanskey, C. A magnetic signature at Io: Initial report from the galileo magnetometer. Sci. 1996, 273, 337-340.

71. McEwen, A.S.; Keszthelyi, L.; Geissler, P.; et al. Active volcanism on Io as seen by galileo ssi. Icarus 1998, 135, 181-219.

72. Stransberry, J.A.; Spencer, J.R.; Howell, R.R.; Diumas, C.; Vakil, D. Violent silicate volcanism on Io in 1996. J. Geophys. Res. 1997, 24, 2455-2458.

73. Veeder, G.J.; Matson, D.L.; Johnson, T.V.; Blarney, D.L.; Goguen, J.D. Io's heat flow from infrared radiometry: 1983-1993. J. Geophys. Res. 1994, 99, 17095-17162.

74. Moore, W.B. Tidal heating and convection in Io. J. Geophys. Res. 2003, 108, 5096-5112.

75. Herndon, J.M. Composition of the deep interior of the earth: Divergent geophysical development with fundamentally

different geophysical implications. Phys. Earth Plan. Inter 1998, 105, 1-4.

76. Lehmann, I. P'. Publ. Int. Geod. Geophys. Union, Assoc. Seismol., Ser. A, Trav. Sci. 1936, 14, 87-115.

77. Birch, F. The transformation of iron at high pressures, and the problem of the earth's magnetism. Am. J. Sci. 1940, 238, 192-211.

78. Ringwood, A.E. Silicon in the metal of enstatite chondrites and some geochemical implications. Geochim. Cosmochim. Acta 1961, 25, 1-13.

79. Fredriksson, K.; Henderson, E.P. Trans. Am. Geophys. Un. 1965, 46, 121.

80. Ramdohr, P. Einiges ueber opakerze im achondriten und enstatitachondriten. Abh. D. Akad. Wiss. Ber., Kl. Chem., Geol., Biol. 1964, 5, 1-20.

81. Ramdohr, P. The opaque minerals in stony meteorites. Elsevier: New York, 1973; p 245.

82. Wai, C.M. The metal phase of horse creek, mount egerton and norton county enstatite meteorites. Mineral Mag. 1970, 37, 905-908.

83. Wasson, J.T.; Wai, C.M. Composition of the metal, schreibersite and perryite of enstatite achondrites and the origin of enstatite chondrites and achondrites Geochim. Cosmochim. Acta 1970, 34, 169-184.

84. Reed, S.J.B. Perryite in the kota-kota and south oman enstatite chondrites. Mineral Mag. 1968, 36, 850-854.

85. Herndon, J.M. The nickel silicide inner core of the earth. Proc. R. Soc. Lond 1979, A368, 495-500.

86. Herndon, J.M. The chemical composition of the interior shells of the earth. Proc. R. Soc. Lond 1980, A372, 149-154.

87. Herndon, J.M. Scientific basis of knowledge on earth's composition. Curr. Sci. 2005, 88, 1034-1037.

88. Herndon, J.M. New indivisible geoscience paradigm. arXiv.org/abs/1107.2149 2011.

89. Murrell, M.T.; Burnett, D.S. Actinide microdistributions in the enstatite meteorites. Geochim. Cosmochim. Acta 1982, 46, 2453-2460.

90. Hollenbach, D.F.; Herndon, J.M. Deep-earth reactor: Nuclear fission, helium, and the geomagnetic field. Proc. Nat. Acad. Sci. USA 2001, 98, 11085-11090.

91. Elsasser, W.M. On the origin of the earth's magnetic field. Phys. Rev. 1939, 55, 489-498.

92. Elsasser, W.M. Induction effects in terrestrial magnetism. Phys. Rev. 1946, 69, 106-116.

93. Elsasser, W.M. The earth's interior and geomagnetism. Revs. Mod. Phys. 1950, 22, 1-35.

94. Lord Rayleigh. On convection currents in a horizontal layer of fluid where the higher temperature is on the under side. Phil. Mag. 1916, 32, 529-546.

95. Raghavan, R.S. Detecting a nuclear fission reactor at the center of the earth. arXiv:hep-ex/0208038 2002.

96. Domogatski, G.; Kopeikin, L.; Mikaelyan, L.; Sinev, V. Neutrino geophysics at baksan i: Possible detection of georeactor antineutrinos. arXiv:hep-ph/0401221 v1 2004.

97. Gando, A.; Gando, Y.; Ichimira, K.; Ikedia, H.; Inoue, K.; Kibe, Y.; Kishimoto, Y.; Koga, M.; Minekawa, Y.; Mitsui, T., et al.

Partial radiogenic heat model for earth revealed by geoneutrino measurements. Nature Geosci. 2011, 4, 647-651.

98. Bellini, G.; et al. Observation of geo-neutrinos. Phys. Lett. 2010, B687, 299-304.

99. Solomon, S.C. Some aspects of core formation in mercury. Icarus 1976, 28, 509-521.

100. Gómez-Pérez, N.; Solomon, S.C. Mercury's weak magnetic field: Result of magnetospheric feedback? Europ. Planet. Sci. Cong. 2010, 5.

101. Stanley, S.; Glatzmaier, G.A. Dynamo models for planets other than earth. Space Sci. Rev. 2010, 152, 617-649.

102. Giampieri, G.; Balogh, A. Mercury's thermoelectric dynamo model revisited. Planet. Space Sci. 2002, 50, 757-762.

103. Eucken, A. Physikalisch-chemische betrachtungen ueber die frueheste entwicklungsgeschichte der erde. Nachr. Akad. Wiss. Goettingen, Math.-Kl. 1944, 1-25.

104. Kuiper, G.P. On the evolution of the protoplanets. Proc. Nat. Acad. Sci. USA 1951, 37, 383-393.

105. Cameron, A.G.W. Formation of the solar nebula. Icarus 1963, 1, 339-342.

106. Wood, B.J.; Walter, M.J.; Wade, J. Accretion of the earth and segregation of its core. Nature 2006, 441, 825.

107. Larimer, J.W. Chemical fractionation in meteorites i, condensation of the elements. Geochim. Cosmochim. Acta 1967, 31, 1215-1238.

108. Urey, H.C.; Craig, H. The composition of stone meteorites and the origin of the meteorites. Geochim. Cosmochim. Acta 1953, 4, 36-82.

109. Herndon, J.M.; Suess, H.E. Can the ordinary chondrites have condensed from a gas phase? Geochim. Cosmochim. Acta 1977, 41, 233-236.

110. Herndon, J.M. Reevaporation of condensed matter during the formation of the solar system. Proc. R. Soc. Lond 1978, A363, 283-288.

111. Daly, L.; Bland, P.A.; Dyl, K.A.; Forman, L.V.; Evans, K.A.; Trimby, P.W.; Moody, S.; Yang, L.; Liu, H.; Ringer, S.P. In situ analysis of refractory metal nuggets in carbonaceous chondrites. Geochimica et Cosmochimica Acta 2017, 216, 61-81.

112. Anders, E.; Grevesse, N. Abundances of the elements: Meteoritic and solar. Geochim. Cosmochim. Acta 1989, 53, 197-214.

113. Fegley Jr., B. High-temperature condensation of iron-rich olivine in the solar nebula. Earth Planet. Sci. Lett. 1987, 82, 180-195.

114. Palme, H.; Fegley Jr., B. Formation of feo-bearing olivines in carbonaceous chondrites by high temperature oxidation in the solar nebula. Lunar Planet. Sci. 1987, XVIII, 754-755.

115. Lodders, K. Alkali elements in the earth's core: Evidence from enstatite meteorites. Meteoritics 1995, 30, 93-101.

116. Herndon, J.M. The object at the centre of the earth. Naturwissenschaften 1982, 69, 34-37.

117. Larimer, J.W. The effect of C/O ratio on the condensation of planetary material. Geochim. Cosmochim. Acta 1975, 39, 389-392.

118. Herndon, J.M.; Suess, H.E. Can enstatite meteorites form from a nebula of solar composition? Geochim. Cosmochim. Acta 1976, 40, 395-399.

119. Lewis, G.N. The law of physico-chemical change. Proc. Am. Acad. Arts and Sci. 1901, 37, 47-70.

120. Herndon, J.M. Evidence contrary to the existing exo-planet migration concept. arXiv.org/astro-ph/0612603 2006.

121. Herndon, J.M. New concept on the origin of petroleum and natural gas deposits. J Petrol Explor Prod Technol 2017, 7, 345-352.

122. Hilgenberg, O.C. Vom wachsenden erdball. Giessmann and Bartsch.: Berlin, 1933.

123. Herndon, J.M. Discovery of fundamental mass ratio relationships of whole-rock chondritic major elements: Implications on ordinary chondrite formation and on planet mercury's composition. Curr. Sci. 2007, 93, 394-398.

124. Reid, A.M.; Bass, M.N.; Fyjita, H.; Kerridge, J.F.; Fredriksson, K. Olivine and pyroxene in the orgueil meteorite. Geochim. Cosmochim. Acta 1970, 34, 1253-1254.

125. Zolensky, M.E.; et al. Mineralogy and petrology of comet 81p/wild 2 nucleus samples. Sci. 2006, 314, 1735-1739.

126. Herndon, J.M. Mercury's protoplanetary mass. arXiv:astro-ph/0410009 1 Oct 2004 2004.

127. Herndon, J.M. Total mass of ordinary chondrite material originally present in the solar system. arXiv: astro-ph 0410242 2004.

128. Herndon, J.M. Hydrogen geysers: Explanation for observed evidence of geologically recent volatile-related activity on mercury's surface. Curr. Sci. 2012, 103, 361-361.

129. Blewett, D.T.; Chabot, N.L.; Denevi, B.W.; Ernst, C.M.; Head, J.W.; Izinberg, N.R.; Murchie, S.L.; Solomon, S.C.; Nittler, L.R.; McCoy, T.J., et al. Hollows on mercury: Messenger evidence for geologically recent volatile-related activity. Science 2011, 333, 1859-1859.

130. http://www.nuclearplanet.com/american%20science%20decline.html

131. http://www.nuclearplanet.com/follies.pdf

132. http://www.nuclearplanet.com/josephson's%20statement.html

133. Herndon, J.M. New concept for internal heat production in hot jupiter exo-planets, thermonuclear ignition of dark galaxies, and the basis for galactic luminous star distributions. Curr. Sci. 2009, 96, 1453-1456.

134. Herndon, J.M. Inseparability of science history and discovery. Hist. Geo Space Sci. 2010, 1, 25-41.

135. http://www.nuclearplanet.com/cv4.pdf

136. Herndon, J.M. Uniqueness of herndon's georeactor: Energy source and production mechanism for earth's magnetic field. arXiv: 0901.4509 2009.

137. Bogue, S.W. Very rapid geomagnetic field change recorded by the partial remagnetization of a lava flow Geophys. Res. Lett. 2010, 37, doi: 10.1029/2010GL044286.

138. Coe, R.S.; Prevot, M. Evidence suggesting extremely rapid field variation during a geomagnetic reversal. Earth Planet. Sci. Lett. 1989, 92, 192-198.